American Indian mtDNA, Y chromosome Genetic Data, and the Peopling of North America

By

Peter N. Jones

The Bäuu Institute

2004

Copyright 2004: Peter N. Jones

The Bäuu Institute

PO Box 4445

Boulder, Co. 80306

All rights reserved. No part of this publication may be reproduced, stored in a retrieval system, or transmitted in any form or by any means, electronic, mechanical, photocopying, recording, or otherwise, without the prior permission of the publisher.

10 9 8 7 6 5 4 3 2 1

Library of Congress Cataloging-in-Publication Data
Jones, Peter N., 1975-
American Indian mtDNA, Y chromosome Genetic Data, and the Peopling of North America.
Includes bibliographical references, index, tables, and images.
p. cm.

ISBN 0-9721349-1-3 (pbk. : alk. Paper)
1. Indians of North America–molecular anthropology–Peopling of the New World. 2. mtDNA–Y chromosome–genetics.

Printed in the United States of America.

ACKNOWLEDGEMENTS

I would like to thank my wife, Tara, for her undivided support and encouragement through the entire creation of this book. I would also like to thank the editors of Bäuu Institute Press for their hard work and dedication to this book.

Table of Contents

Introduction	1
Molecular Anthropology: A Review of the Field	3
Concepts and Terminology	7
Haplotypes and Haplogroups	15
Coalescent Trees and Gene Trees	19
Molecular Anthropological Population Theory	27
Human Migrations and Molecular Anthropology	33
American Indian mtDNA and Y chromosome Studies	47
Summary and Conclusion	67
References	75
Appendix	107
Index	219

CHAPTER ONE
INTRODUCTION

Over the past two decades physical anthropologists and molecular geneticists have begun to use molecular data in their attempt to answer several, long standing questions, especially those of migrations in human history. This new line of research has been called genetic anthropology, molecular anthropology, and archaeogenetics (see Renfrew & Boyle, 2000). The term molecular anthropology best expresses the purpose and direction of this relatively new field.[1] Presently most of the information related to this field has come from DNA (deoxyribonucleic acid) obtained from living human populations, though there are a handful of studies using ancient DNA (aDNA), molecular data recovered from historic bones, teeth, and other skeletal remains. Furthermore, within this line of research, many new procedures and models have been used, some more robust than others, which have allowed researchers to yield conclusions about the relationships between populations, both past and present. Therefore, as this book will discuss, researchers have been able to hypothesize about relationships between present populations and their demographic histories, which are situated in the past.

This book is a comprehensive analysis of the history, theory, and current state of the field of molecular anthropology, focusing in particular on the use of mitochondrial DNA (mtDNA) and Y chromosome genetic material of North American Indians and the peopling of North America. The primary reason for this book, besides the need for an academic review of the field for the general anthropological, legal, and cultural heritage community, is that molecular anthropology is currently given more scientific weight within the larger scientific and legal communities than other epistemological forms such as oral traditions. In brief, this book begins with a brief history of the development of this new line of research, followed by a section covering the relevant terminology and conceptions employed within the molecular anthropological field. Penultimately, two sections discussing the major findings of the field are offered, the first focusing on mtDNA research, and the second discussing Y chromosome research. Finally, the book ends with a summary and conclusion discussing how this data fits into the larger field of anthropology and the discussion of the peopling of North America. An appendix is also included, abstracting all studies that were located which have been conducted concerning the issues addressed in this book, as well as giving the title of the study, authors, citation, and genetic materials used.

FOOTNOTE

[1] I claim that this is a relatively new field because the theories, methods, and techniques employed in the study, use, and analysis of mtDNA and Y Chromosome data draws more from biology and genetics then from historically based physical anthropology. Therefore, in the tradition of Kant in designating philosophical anthropology, molecular anthropology has so been recognized within the last couple of decades as a legitimate subfield of physical anthropology.

CHAPTER TWO
MOLECULAR ANTHROPOLOGY: A REVIEW OF THE FIELD

As briefly stated in chapter one, molecular anthropology as a field deserving its own name is only a couple decades old. The first attempt at using genetic data to investigate anthropological questions was published in 1965 by Cavalli-Sforza and Edwards, entitled *Analysis of human evolution*. This pioneering paper reconstructed historic population movements (also referred to in the literature as demographic histories, prehistoric or ancient migrations, prehistoric or ancient population affiliations, or phylogenetic relationships; the term historic population movement will be used in this book) on the basis of "classical" genetic data based on samples taken from living populations. Classical genetic data uses proteins and blood groups, as opposed to molecular genetic studies that use mtDNA and Y chromosome data. Cavalli-Sforza, along with Menozzi, and Piazza subsequently went on to compile their magisterial *The History and Geography of Human Genes* (1994), which relied primarily upon classical genetic markers sampled on a worldwide basis. This book has been taken as a marker of the emergence of molecular anthropology as a distinct subfield within anthropology, a time when the use of such classical genetic markers was replaced by molecular studies.

The "birth" of molecular anthropology proper, which this book is concerned about, is currently in full spate, and was initiated by the earliest papers utilizing DNA sequencing for the reconstruction of human population histories. One of the first of these, entitled *Evolutionary Relationships of Human Populations from an Analysis of Nuclear DNA Polymorphisms* (Wainscoat et al., 1986) used nuclear DNA.

The important paper by Cann, Stoneking, and Wilson (1987) entitled *Mitochondrial DNA and Human Evolution,* was one of the first to utilize the potential of mitochondrial DNA for studying specific lineages of the female line. These studies focused on the larger questions of human evolution, and not necessarily the demographic history of North American Indians.

More recent studies, however, have focused extensively on North American Indian demographic histories and the peopling of North America using both mtDNA and Y chromosome data. These studies have used genetic data to claim a link between European/Western Asian populations and North American Indian populations; both populations having recent common ancestry (Brown et al., 1998). Other studies have claimed to have identified a single wave of migration for the peopling of the Americas (Bianchi et al., 1997; Bianchi et al., 1998; Easton, Merriwether, Crews, & Ferrell, 1996; Merriwether & Ferrell, 1996; Merriwether, Hell, Vahlne, & Ferrell, 1996; Merriwether, Rothhammer, & Ferrell, 1995) as opposed to several (Karafet et al., 1997; Karafet et al., 1999), while a few studies have concluded that some American Indian tribes have recently moved into specific geographic areas (Kaestle, 1995; 1997; 1998; Kaestle & Smith, 2001b), despite contrary evidence from oral history and archaeology. Some of the most publicized uses of molecular anthropology in recent years concerns the question of biological affiliation, as part of compliance procedures associated with the Native American Graves Protection and Repatriation Act (NAGPRA; 25 USC 3001 et seq.; 43 CFR 10), exemplified in such cases as the Spirit Cave Mummy and the Kennewick Man repatriation controversies (Kaestle, 2000a; 2000b; Merriwether & Cabana, 2000; Merriwether, 2000; Tuross & Kolman, 2000). In these two examples the situation is complicated by the great antiquity of the skeletons, 9,415+/-25 years ago and 8410+/-60 years ago, respectively (Napton 1997; Chatters 2000).

This new line of research, however, is not only concerned with migrations. In fact, the potential benefits from this new line of research offers to be vast and highly valuable.

Such potential benefits include a better understanding of the genetic and evolutionary factors that influence populations; an understanding of maternally transmitted diseases such as blindness, epilepsy, dementias, cardiac and skeletal muscle diseases, diabetes mellitus, and movement disorders; the development of new metabolic and genetic therapies for mitochondrial diseases; and a better understanding of the geographic origin of anatomically modern humans, to name just a few. However, these studies are still rather rare because of the lack of information, partially resolved with the sequencing of the human genome. This is an area of research that molecular and medical anthropologists could collaborate on, to the benefit of both science and humanity.

Instead, most studies have used the new molecular genetic data to investigate questions of biological affiliation and historic population movements. The earliest studies identified to use either mtDNA or Y chromosome material in exploring the question of historical population movements are those of Aquadro and Greenberg (1983), Johnson, Wallace, Ferris, Rattazzi, and Cavalli-Sforza (1983), and Salzano (1982). Erdtmann, Salzano, and Mattevi published an early paper discussing the use of Y chromosome data and South American Indians (1981), but the first studies to discuss North American Indians were not done until the mid-1980s (Paabo, Gifford, & Wilson, 1988; Wallace, Garrison, & Knowler, 1985). For this book 96 studies were located and analyzed, though there is some possibility that others exist, especially in obscure or regional journals or as technical reports not readily available to the scientific community.

CHAPTER THREE
CONCEPTS AND TERMINOLOGY

As mentioned in the first chapter, this new line of research draws on the fields of anthropology and molecular genetics, and therefore, several terms are used that may not be familiar to many. Therefore, this section covers the predominate concepts and terminology found within the literature.

As most everyone knows, we all carry deoxyribonucleic acid (DNA) in every cell of our bodies, which has been passed down almost unchanged from our earliest ancestors. DNA, therefore, is the messenger of heredity. A simple metaphor to help explain much of the terminology used in molecular anthropology is that of an instruction manual. We can view DNA as a set of written instructions on how to build a human with the chromosomes acting as volumes of the manual. Not surprisingly, these instructions are immensely complicated and nowhere near fully understood. Nonetheless, the language of the instructions is very straightforward. Like many languages, the meaning is contained within a sequence of symbols or letters, of which the genetic language contains four symbols. These four symbols are the simple organic chemicals adenine, cytidine, guanine, and thymidine, always referred to as A, C, G, and T. These four chemicals, the nucleotide bases, are joined together one after another in a long molecular chain that forms DNA. In fact, the DNA molecule consists of two strands, the famous double-helix, each one containing the same information in its sequence of bases but in a complementary way. Therefore, when A appears in one strand, it is always opposite a T in the other. G and C are similarly matched.

When cells divide, DNA must be copied so that each daughter cell receives a full set of instructions. This is accomplished by unwinding the double-helix, and using each single strand as a template, to make two new identical double-helices. Because of the complementarity of the bases, the sequence remains intact. The copying mechanism (mitosis) is remarkably exact, but there are occasional mistakes, called mutations. It is these mutations, introduced randomly into the DNA molecule, that molecular anthropologists look for to compare.

Molecular anthropology, therefore, doesn't compare the blood from one individual to that of another, as "classical" genetics and physical anthropology does, but instead compares polymorphic genetic frequencies to those of others. A polymorphism, as the name implies (i.e., many forms) is the condition that within a population there exist differences in the population genetic structure, based on mutations. This implies the presence of two or more alleles – actual alternative variants that are similar but not identical – located at a particular position or locus on a chromosome. The human genome – the collective name for all DNA in each cell – is organized into what are called chromosomes, separate "volumes" of the human genome that reside in the cell nucleus. All of the chromosomes contain the three billion symbols that make up the human genome. In all there are twenty-four different chromosomes in the human genome; twenty-two of which come from the parents (eleven from the mother, eleven from the father). These twenty-two chromosomes are collectively known as the autosomes, which distinguish them from the X and the Y sex chromosomes (chromosomes 23 and 24). Females have a pair of X chromosomes while males have both an X and a Y chromosome.

The classic biochemical approach, and the one used by molecular anthropologists, for investigating historic population movements consists in taking samples, usually blood samples, from a well-defined human population and testing these to determine the presence or absence of alleles for the given polymorphisms under investigation. The number of individuals within the given sample of the population who possess a particular allele is then expressed as a gene frequency.

These molecular frequencies are called haplotypes, and several haplotypes (i.e., gene frequency varieties) make up a haplogroup.

By documenting the various mutations found in a population (the population's molecular frequency for a particular allele) and comparing these molecular frequencies to those of other populations, molecular anthropologists can begin to reconstruct the historic population movements of the two populations under study. This is usually done using theories modeling coalescent times and divergence times, which will be explained in further detail below. Through this process the analysis of the spatial- and community-specific distribution of a set of haplogroups has led to the development of a new field in evolutionary biology: phylogeography (Advise, Arnold, & Ball, 1987), though as discussed in this book the term molecular anthropology is used.

Following the construction of intraspecific molecular phylogenies, a molecular anthropological analysis allows the distribution of gene genealogies to be traced in space and time. Assessing the coalescence of lineage ancestry within, and among, populations, therefore, provides information about the origin of contemporary genetic variation. In particular, the distribution of haplogroups can be used to draw conclusions about the relative impact of deterministic forces compared to the influence of genetic drift. Molecular anthropological analyses can also be used to investigate the influence of past demographic fluctuations (Rogers, Rogers, & Martin, 1992; Slatkin & Hudson, 1991). However, within this relatively novel area, the intersection between theory and observation still needs considerable development.

The human genome, along with the 24 chromosomes, contains one other piece of DNA, which is contained not in the nucleus but in small particles in the cell cytoplasm, called mitochondrial DNA (mtDNA). It is much smaller than the nuclear genome with just over 16,500 bases compared with the 3000 million bases found in the nuclear DNA. MtDNAs peculiar genetic characteristics, however, have made it a central component of molecular anthropological research.

As discussed above, there are several elements that make mtDNA particularly useful in the study of historic population movements. The first is that unlike nuclear DNA it is inherited from only one parent – the mother. This is because human eggs have a large cytoplasm full of mitochondria while human sperm contains only a few, and those few either do not get into the fertilized egg or are eliminated shortly afterwards. This has two major implications. First, this means that all people inherit all of their mtDNA from their mother, who inherited her mtDNA from her mother, and so forth. Therefore, at any time in the past only one woman was an individual's maternal and hence mtDNA ancestor. The other important implication is that mtDNA does not undergo genetic recombination. Recombination is the device used by chromosomes to shuffle their genes at every generation, which has the evolutionary advantage that new, favorable gene combinations occasionally emerge. These two features have proved useful because it has allowed molecular anthropologists the ability to track the rare mutations that arise between generations of the mtDNA, thus allowing them to document mtDNA allele frequencies of particular maternal lines. It is these allele frequencies that can then be compared to other allele frequencies to calculate when the two maternal lines historically diverged.

The Y chromosome also shares these two features of mtDNA, namely uniparental inheritance and a lack of recombination. Molecular anthropologists use variations that arise through mutations in both mtDNA and Y chromosome data to trace populations. Most mutations arise during the DNA-copying process prior to cell division. Of these possible mutations the simplest type of mutation is known as a point mutation, where the replacement of one base (A, C, G, or T) is replaced by another. This always happens in one individual cell in one individual person. To be passed on to the next generation the mutations must occur in the so-called germ line cells that are the precursors of either eggs or sperm. All sorts of mutations occur in other body cells, but these are irrelevant to the study of historic population movements, because they do not get passed on to the next generation. Furthermore, the mutations will have to increase to be noticed at all. If the new mutation does not alter the biological fitness of the individuals carrying it, in other words if it is a neutral change, then the process by which it spreads, or is eliminated, is governed purely by chance and referred to as genetic drift. Therefore, taking the Y chromosome as an example, suppose a mutation happened in the germ line of a male. If he did not have any sons then the new mutation would not be passed on. However, if a mutation arises in a males body cells, it does not matter if he has sons or not, because body cells are not transmitted from generation to generation. MtDNA is more complicated, however, since a new mutation arising in a female germ line will only be in one molecule to begin with. There are thousands of mtDNA in each cell, and along the line of cell divisions to the mature egg cell the number of mitochondria are successfully reduced to one, so that there is a chance that this new mutation might not get passed on to the next generation. If, however, the mutated mtDNA cell does slip through the cellular "bottleneck" and get into the next generation, the new allele might manage to reach a reasonable proportion in the egg cells sufficient to be noticed in, say a blood sample of the individual. This transition state, between detectability and undetectability, is known as heteroplasmy and persists for half a dozen generations or so before the new mutation either takes over the entire germ line (fixation) or recedes into oblivion. Even when fixed, though, the new version is far from secure because if the women who carry it do not have any daughters then it will not be passed on any further.

As mentioned, single point mutations are the commonest source of variation in mtDNA. The circular mitochondrial genome contains the genes for making the components of aerobic metabolism, leaving about 1000 bases that do not code for anything (that researchers have found yet). Mitochondria, because they do not recombine, are not able to correct for copying errors, which has led researchers to theorize that mtDNA undergoes a constant mutation rate, estimated at 20 times faster than that of nuclear DNA. Furthermore, the 1000 bases of non-coding mtDNA – the so-called control region

(CR) – accumulates mutations even faster than the rest of the mtDNA, making it by far the most variable stretch of DNA in the whole human genome. Sequencing this 1000 base segment, or even just two segments within it, called hypervariable segments I and II (HVS I and II), has proven to be a very productive way of researching mtDNA variation.

There are other variable bases outside the control region and a selection of these is often recruited to clarify the control region variation. Because these mutations are rather thinly spread they tend not to be sequenced directly, but detected by their ability to create or destroy so-called restriction sites. Restriction sites are short DNA sequences, typically 4 - 6 bases in length. These areas of the genome are referred to as Restriction Fragment Length Polymorphisms (RFLP), and are also used in the study of historic population movements.

The Y chromosome also has, spread along its length, a useful selection of RFLPs, now often referred to as the bi-allelic markers, so named because they distinguish two alleles, one where the restriction site is intact and the other where the site is disrupted. There is also another useful source of variation in the Y chromosome, called the microsatellites. For some reason certain very short DNA sequences of 2, 3, or 4 nucleotides in length have a tendency to grow. The short sequence is repeated several, sometimes hundreds, of times. In some, but not all, the number of repeats is unstable and several versions of different lengths are to be found. This has proven to be an excellent source of variation, especially when combined with the bi-allelic data from the same individual. Complicating these factors is that mutational events must be distinguishable, a critical requirement for estimating a phylogeny for a specific genomic region. Although this is well-nigh impossible for microsatellites, the development of the minisatellite variant repeat mapping strategy (Jeffreys et al 1991) holds promise for minisatellites. This technique may prove exceptionally useful for defining phylogenies for otherwise uninformative regions, such as the Y chromosome. With this technique, the ancestral state of mobile elements (e.g. Alu elements) can be defined; therefore, these loci hold considerable promise for developing population phylogenies (Batzer, Stoneking, & Alegria-Hartman, 1994). However, individual elements are relatively uninformative for a single genomic region. The incorporation of multiple loci into haplotypes also holds promise, but more data are needed on the relative rate of recombination versus mutation. Ultimately, the sine qua non for phylogenetic reconstruction is sequence data.

It is likely that most mutations in microsatellite loci involve a change in size of one or two repeat units, presumably by slippage. Infrequent events cause changes in size of several repeat units. Although such large changes in repeat number could also occur by slippage, the evidence for this is less clear than it is for small changes, and alternative models have been proposed, such as hairpin formation. Direct sequence data indicate that many "simple repeat sequences" are not so simple and that mutations occur through alterations in base composition. In particular, the conversion of cryptic to perfect repeats is responsible for an increased mutation rate of microsatellites, including the unstable expansions observed in some trinucleotide repeats associated with inherited human diseases. These observations suggest that mutation rates may differ substantially between different alleles at a single locus; so far this hypothesis has only been tested for loci with unusually high mutation rates (Freimer & Slatkin, 1996).

Variable minisatellites, tandemly repeated units often much longer than microsatellites and frequently with a complex internal structure, are found more rarely, but when they are, such as MSY1 located on the Y chromosome, they can also be useful. The study of these regions on the mtDNA and Y chromosome and their allele frequencies are called haplotypes. Many haplotypes, as mentioned, make up a haplogroup, and it is this larger entity that is used to compare one population with another.

CHAPTER FOUR
HAPLOTYPES AND HAPLOGROUPS

Although researchers have noted that limitations exist when studying only one gene (Chen et al., 2000; Karafet, Zegura, Vuturo-Brady, Posukh, Osipova, Wiebe, Romero, Long, Harihara, Jin, Dashnyam, Gerelsaikhan, Keiichi, & Hammer, 1997; Mountain & Cavalli-Sforza, 1997), most molecular anthropological studies still rely on only one gene and its alleles because of the ease in identifying differences in a restricted location on that gene, especially in non-recombining genes such as mtDNA and Y chromosome. The allele sequences that are studied are called haplotypes, which for American Indians presently fall into five larger recognized haplogroups (A, B, C, D, and X), and have been used in most studies concerning American Indian population genetics.

It is necessary to point out several assumptions underlying the uses of haplotypes and haplogroups for their use in the literature is ubiquitous. First, many studies use within-local-population frequencies for the genetic sequences, which are highly affected by each population's specific recent demographic history, and the possibility exists that researchers will underestimate the nucleotide diversity of the population as a whole (Bonatto & Salzano, 1997a). Therefore, the differing results between CR (control region) sequences and RFLP (restriction fragment length polymorphism) sequence data cannot be explained either by sample size or attributed to the different ways in which the haplogroup frequencies were treated, but are more probably due to the different populations or regions of the DNA studied.

Likewise, as previously noted, the only changes introduced into genes are either point mutations, insertions, and/or deletions (with insertions and deletions being rare in comparison to point mutations). This means that each possible founding lineage cluster can be thought of as containing the founding lineage haplotype plus a collection of that lineage's descendants. However, there are several problems inherent in this assumption, notably that the original Y chromosome can eventually die out, shifting time, haplotype frequency, or relationships of the population under study (Bradman & Thomas, 1998), and can result in faulty data when comparing a present population's frequencies to those of an ancient population's haplotype frequencies (whether actually based on aDNA or hypothetical frequencies). As Bradman and Thomas (1998) pointed out using the insertion of the YAP (Y chromosome alu polymorphism) indel (insert) on the Y chromosome, descendents of individuals after only one generation may not carry the same Y chromosome alleles. It is possible that a descendent of the individual who first acquired the YAP indel may lose that indel, yet still remain a descendent of that individual. This is also possible with mtDNA, where a father's son or daughter will not carry the genetic information of that person's father's mother. By only looking at specific alleles, mutations, insertions, and deletions can be viewed as coming from discontinuous populations. Furthermore, as Bianchi et al. have pointed out, "the combination of a decrease in the effective population size and genetic hitch-hiking may have been the cause producing a single variety of Y-chromosomes in the earliest ancestors of extant Amerindians" (1997, p. 87). If this is correct, then spurious results may arise when determining biological affiliation between populations.

Finally, as noted, the mitochondrial genome undergoes no recombination, and therefore the 16,569-bp genome behaves evolutionarily as a single locus. As MacEachern (2000, p. 358) noted, "In particular, it appears that there may be significant variability in selection mechanisms on the genome itself and in the mitochondria and in rates of phylogenetic versus intergenerational mtDNA mutation that are only now being appreciated (Gibbons 1998; Parsons, Muniec, and Sullivan 1997)." Therefore, inferences from any one such locus lack robustness (Pamilo & Nei, 1998).

The patterns observed in genetic data from within populations reflect the effects of mutation at the locus in question superimposed on the genealogy of the sampled genes. Although it is not directly observed, the Most Recent Common Ancestor (MRCA) of the sampled genes will be of a particular allelic type. In the absence of mutation, all the sampled genes would be of this type. The variation actually present in the sample results from mutations along the lineages leading down from the MRCA to the sample.

A mutation that occurs on a particular branch of the genealogical tree will affect all the genes in the sample, which are descended from the ancestral gene associated with that branch. For example, under the "infinite sites" assumption that no back mutation has occurred between the MRCA and the sampled genes, each mutation on the tree will correspond to a segregating site in the sample. For a particular such mutation, all the genes descended from the ancestral gene that underwent mutation will have one base at a particular site, whereas all the other genes will have a different base, that of the MRCA, at that site. Mutations that occur higher up the tree, closer to the MRCA, will tend to be represented in more of the sampled genes.

CHAPTER FIVE
COALESCENT TREES AND GENE TREES

Once the haplotypes of a population have been determined, a unique gene tree can be constructed from the configuration of mutations under the assumption that point mutations arise at sites only once in time, without any back or parallel mutation. The coalescent tree is hypothetically a perfect phylogeny representing the mutation history of that haplotype back in time. The coalescent tree is equivalent to the DNA sequence data, and because it hypothetically represents the ancestry of the population, it is common to think of the DNA sequence data as a phylogenetic tree. It is important to remember, however, that the data is not independent of the relationships shared through ancestry. The likelihood of a coalescent tree under a stochastic coalescent model of evolution can be found by advanced simulation techniques, thus allowing a maximum likelihood estimation of the parameters using the full information in the data. The distribution of the time to the most recent common ancestor and ages of mutations in the tree, conditional on its typology, can also be found by simulation techniques. Computing likelihoods by computer intensive methods for samples of DNA sequences under general models is currently a very active research area. Some of the approaches used are Importance Sampling (Bahlo & Griffiths, 2000; Fearnhead & Donnelly, 2001; Griffiths & Marjoram, 1996; Griffiths & Tavare, 1994a; 1994b; 1994c; Nielsen, 1997; Slade, 2000a; 2000b); Markov Chain Monte Carlo (MCMC) by Felsenstein and colleagues (1995; 1998; 1997); and other approaches, such as MCMC of a Bayesian nature by Wilson and Balding (1998), Beaumont (1999), and Markovtsova et al. (2000a; 2000b).

One important quantity in classical population genetics is the inbreeding coefficient, a measure of mating between relatives, that is defined as the probability that a pair of genes at a locus are identical by descent. A pair of genes are considered identical by descent if both are derived from the same gene in a common ancestor (Crow & Kimura, 1970), and Gustave Malecót (1955) was one of the first who clearly distinguished this concept from identity-in-state. The inbreeding coefficient is intimately related to the effective number of individuals in a population that is the size of an ideally behaving population that would have the same decrease in heterozygosity as the observed population. The effective number or effective population size is used if there are fluctuations in the population number from time to time, or if the distribution of number of progeny per parent is nonbinomial, or if there is any other kind of deviation from the idealized model that has been assumed (Crow & Kimura, 1970). Therefore, in a hypothetically random mating population of effective number N_e, the inbreeding coefficient is $1/(2N_e)$ at a diploid locus. Malecót (1955; 1967) showed that the expected heterozygosity decreases in time with this rate of inbreeding coefficient and that in a randomly mating population of N_1 males and N_2 females, N_e is given by $4N_1N_2/(N_1 + N_2)$. This theory has been the basis for the coalescence theory of modern population genetics (Kingman, 1982).

Furthermore, under neutrality (Kimura, 1968), the number of polymorphisms are determined by the effective size of the population (N_e) and the hypothetical neutral mutation rate. The nucleotide diversity (Nei, 1987), and the number of segregating sites (Watterson, 1975), are the commonly used measures for DNA polymorphism in such a model. Since both measures are simply related to a single population parameter ($4N_e g ?$, where g is the generation time and ? is the neutral mutation rate per site per year), it is possible to estimate N_e from observed values of the nucleotide diversity and/or the number of segregating sites. It is important to point out that this model assumes both non-random mating and constant mutation rates, two features that do not occur in human populations and therefore can lead to spurious results. Likewise, neither the nucleotide diversity nor the number of segregating sites is observable for an ancestral species, which is essential if one is attempting to research historic population movements. They, and any other quantities, must be inferred indirectly, and precisely because of this, two methods have been proposed.

The first method, called the trichotomy method, uses gene genealogies among three species that diverged from each other in close succession. This method can be used in questions of hominid origins, comparing chimpanzees (Pan troglodytes), gorillas (Gorilla gorilla), and human (Homo sapien sapien) data to see when the three species split. Obviously, this method is not applicable for the study of prehistoric human population movements, and will not be discussed.

The second method, called maximum likelihood, uses pairs of orthologous sequences sampled from two species only (Takahata & Satta, 1997; Takahata, Satta, & Klein, 1995). This is the method preferred in studying human populations, and in studying biological affiliation between two populations the same principles have been assumed valid, i.e., orthologous sequences sampled from two different populations (as opposed to two different species). Such orthologous sequences, however, must have diverged prior to the populations splitting, thus it is important to correlate the two populations under comparison with corroborating evidence that will support that the two populations were at one time genetically related. This corroborating evidence can either be archaeological or linguistic in nature. While segregation occured in the ancestral population, it is hypothesized that two orthologous sequences developed and accumulated nucleotide substitutions to form an ancestral polymorphism. When the subsequent populations split, it is hypothesized that this also allowed them to further differentiate by population-specific nucleotide substitutions. It is assumed, therefore, that the number of nucleotide substitutions per sequence prior to and posterior to population splitting follows geometric and Poisson distributions (the number of events occurring within a given time interval), respectively. The principle of this method is to separate these two types of substitutions when a number of independent pairs of orthologous sequences are available. One important assumption is that the neutral mutation or substitution rate is kept constant over the nucleotide sites under study. Hence, only synonymous sites, or introns in coding regions, and intergenic regions are preferably used. However, as already mentioned, researchers are not sure whether nucleotide sites undergo a constant mutation or substitution rate.

Researchers, because of this lack of assurance, have concluded that if intragenic recombination is frequent it will lead to erroneous estimates of ancestral polymorphism. Two incompatible requirements, therefore, arise. To infer accurate gene genealogy, researchers must look at long stretches of DNA in which a sufficiently large number of nucleotide differences can be observed. On the other hand, such long stretches are likely to undergo intragenic recombination, resulting in faulty genealogies.

Likewise, it is essential that the researcher, in using genetic frequencies and coalescent times, does not assume that these are the same as the times of origin for the population under study or when one population split from another (i.e., biological affiliation). Although tracing the genealogy of mtDNA or Y chromosome allele frequencies theoretically can lead to a single common ancestor, this is not evidence that the populations under study went through a period when only one ancestral breeding population was alive and reproducing. Tracing the coalescent times leads to one ancestor of a unilineally transmitted set of markers (either through the maternal or paternal line), but the descendents of the original DNA will have had haplotype frequencies that differed among that of the entire population, resulting in a biased sample of the total historic population's frequencies when using coalescent times. This is because working back in time does not allow one to take into account the various branches of diversity that the historic population had, but only the lineal history of the specific marker being coalesced. Three primary assumptions arising from the use of coalescent times (Hoelzer, Wallman, & Melnick, 1998; Hudson, 1990; Templeton, 1993; 1998; 2002) that have been employed specifically in understanding

American Indian historic population movements are:

1) gene coalescence is a regular process of mutation accumulation in neutral systems, and therefore can be timed like a regularly ticking clock with an acceptable range of error;

2) American Indian populations were isolated from each other after they originated or migrated to the Americas; and

3) the history of particular gene systems is the history of the specific populations in which they are found.

However, as already mentioned, human populations are not neutral systems, and it is not clear if it is safe to assume that mutations occur in a regular, timely fashion. Furthermore, as much of the American Indian ethnographic, linguistic, and archaeological data demonstrates, American Indian populations were never isolated, either from each other or possibly from ancestral populations in Asia (for discussions on the latter aspect, see Akazawa, 1999; Anderson & Gillam, 2001; Bever, 2001; Ikawa-Smith, 1982; Tarazona-Santos & Santos, 2002).

One important requirement in coalescence theory is the use of random samples of genes from the population under study. However, this is extremely difficult to accomplish, not to mention when studying historical relationships between ancient populations and their possible descendents. As Donnelly and Tavare (1995) point out,

> In practice, genetic data are typically obtained from convenience samples rather than proper random samples. There is an obvious danger that such data may contain individuals who share relatively too much ancestry on the relevant timescales. The extent to which application of

coalescent (or traditional) methods to such convenience samples may be misleading remains an open, and potentially serious, question. (p. 418)

Furthermore, most studies that research American Indian historic population movements rely on the idea that American Indians came to the Americas from Asia in small groups (usually thought to have occurred as part of one to three migration waves) across the Bering Land Bridge in prehistoric times. If this is the case, coalescence times will be shorter than actual population divergence times because smaller populations in the past are more likely to share ancestors (Donnelly & Tavare, 1995, p. 410), leading to an accelerated time of origin for American Indians, and thus not correctly representing occupational time depth or biological affiliation.

Similarly, departures from random mating due to inbreeding, assortative mating, or population stratification can lead to non-random association between genotypes and further complicate the interpretation of the data and coalescent times. As Karafet et al. (1997) concluded, because of the presence of the 1T haplotype (a Y chromosome combination haplotype) in both northeastern Siberia and the Americas, the possibility of historic and prehistoric back-migration is extremely likely. Similar studies have also noted the possibility of gene transfer or the "hitch-hiking theory" among American Indian and Asian populations (Bianchi, Bailliet, Bravi, Carnese, Rothhammer, Martinez-Marignac, & Pena, 1997; Bradman & Thomas, 1998; Hudson, 1990). Because population-coalescence times are frequently a result of the fusion of several of the ancient phylogenetic clusters and not necessarily the age of individual populations (Watson, Forster, Richards, & Bandelt, 1997), faulty results may be reported. Therefore, using gene coalescent times as possible times of origin for American Indians can lead to spurious conclusions, for there is no evidence that American Indians were ever: 1) part of a neutral system that can be timed like a regularly clicking clock, 2) were isolated from each other or from Asian populations, and 3) that the current gene systems found in a particular population fully represent the historical diversity of that population. The data for

Amerindian tribes suggest that bands (or villages) are fairly ephemeral with short coalescent times resulting in a high degree of lineage sharing. Analysis of lineage frequency, rather than phylogeographic structure, is most likely to be informative. However, tribes appear to be more stable entities, with considerable scope for phylogeographic analysis.

CHAPTER SIX
MOLECULAR ANTHROPOLOGICAL
POPULATION THEORY

The mtDNA and Y chromosome sections of the human genome have proven to be the most useful for studying historical population movements because of their ease in replication and amplification, as well as the fact that they are non-recombining. However, the larger theoretical assumptions underlying how molecular anthropologists reconstruct particular population allele frequencies is still nascent. In 1985, Jeffreys et al. (1985) introduced individual-specific "fingerprints" for multiple loci, which were later applied to single-locus variable numbers of tandem repeat (VNTR) polymorphisms and short tandem repeats (STRs). In parallel with these molecular advances, which made DNA typing more sensitive and reliable, the mathematical theory became more precise. In rapid succession three main obstacles were overcome: population structure, kinship, and database trawls. All depend on probability theory, the first two deriving from Malecot's work (1955; 1967; 1973; 1975), which demonstrated that weight of evidence is measured by the likelihood ratio (LR):

$$LR = \frac{P(E_c | E_s, H_1)}{P(E_c | E_s, H_0)} \equiv \lambda$$

where E denotes mtDNA or Y chromosome evidence, C is the genetic mtDNA or Y chromosome frequency within that population (already discussed above), and S is a sample from the skeleton, bone sample, or individual in question. The null hypothesis H_0 is that S is not equal to C (different blood lines; no biological affiliation) and the alternative hypothesis is that S = C (the same blood line; biological affiliation). These probabilities are functions of gene frequencies, at least one parameter of population structure ?, and perhaps N, and the number of individuals in a database trawl. Certain conditions for mtDNA or Y chromosome identification are outside the probability theory, which also must be taken into account in genetic affiliation studies. For example, the "chain of custody" must be preserved from the location where the sample was taken through testing, with adequate guarantees against tampering, misidentification, and contamination. Furthermore, there must also be guarantees that testing was performed without error, and probabilities under a given hypothesis are evaluated correctly. Likewise, gene frequencies must be estimated from appropriate random samples of the relevant population and that other parameters are appropriate. Bayesian methods that specify prior probabilities for the various assumptions and hypothesis are too subjective to be considered in presenting the evidence.

A complicating factor in this line of research is that genotypes are not unique: monozygotic cotwins have the same genotype, and any other individual (whether related or not) has a finite probability of the same genotype at a finite number of loci. This means that DNA evidence must be presented in terms of matching probabilities.

It is for the above reasons that adequate sample sizes of the populations under study be used. Variations in population size are commonly attributed to bottlenecks and the so-called founder principle, in which a population encounters a severe reduction in size or a few individuals colonize a new area, resulting in a small selection of gene frequencies as compared with the original population. However, an important complication that makes it impossible to determine census size of a prehistoric human group as a direct estimate of the effective population size is that human populations have overlapping generations. Rogers and Jorde (1995) have shown that the only sense in which sequence diversity can be employed as a measure of chronological age is as an estimation of the time during which a particular population has expanded after experiencing a severe bottleneck. This is because we are dealing with allele frequencies (haplotypes), and not with distinct populations. In fact, the error variance increases with time and the earliest observations are the most precise. For example, computer simulations that suggest that the four major haplogroups found among American Indians underwent a bottleneck followed by a large population expansion may be questioned. These simulations are based primarily on the analysis of CR sequences from haplogroup A and do not take into account haplogroups B, C, D, and X. Similarly, although most studies investigating human population movements have used sequence diversity as a measure of age, few have investigated whether their samples met the very stringent assumptions required by this practice (Bonatto & Salzano, 1997a, p. 1417). Furthermore, Bonatto and Salzano (1997) have also noted that studies using RFLPs have found that haplogroup B has a much lower diversity than the other four (A, C, D, X), which would lead to inaccurate computer simulations. Therefore, for example, the current dates from mtDNA and Y chromosome studies contending that American Indians arrived in the "New World" around 35,000 years ago can be questioned (Bonatto & Salzano, 1997a; Bonatto & Salzano, 1997b; Brown, Hosseini, Torroni, Bandelt, Allen, Schurr, Scozzari, Cruciani, & Wallace, 1998). This number is actually the time during which American Indians theoretically experienced an expansion after a bottleneck. However, it is unknown if this bottleneck took place in Asia, the generally accepted place of origin of American Indians, or in the Americas after their arrival, nor is it known what effects subsequent migrations and bottlenecks from disease and other factors have on this time estimation.

Adequate sample sizes are also critical if the genetic frequencies used to characterize a population are to be considered reliable. Typically in studies addressing American Indian historic population movements, sample sizes range between four and 30 individuals per tribal population; this is insufficient to detect little more than the most common haplotypes in each population. Although it is necessary to have genetic samples from 50 males or 50 females of an individual population to accurately infer genetic demographic history, very few studies have done this. The largest study to date on American Indians dealt with 2,198 males from 60 global populations, including 20 American Indian groups (Karafet et al. 1999; this study relied on large amounts of data gathered from previously published reports, and thus could not correct for those sample sizes). However, only the Inuit Eskimo and Navajo samples were over 50 at 62 males and 56 males, respectively. All others ranged from as high as 44 to as low as two individuals. It is unrealistic to assume that one can get an accurate picture of a tribe's genetic frequencies using only two males. In fact, Weiss (1994, p. 834) suggests that we may not be able to distinguish loss of lineages after one migration or separate migrations from a common source population, thus further stressing the critical need for adequate population sample sizes. As Ward et al. (1993) have noted, a sample size of 25 will detect ~63 percent of the lineages in a tribe with normal diversity. In tribes with extensive diversity a sample size of 25 individuals will only detect ~40 percent of the lineages and sample sizes of 70 or above are required to detect two-thirds of the lineages. The fact that the majority of studies lack the required sample sizes necessary to detect even 63 percent of the lineages in a normally diverse tribe brings into question many of the results of these studies, especially when it has been noted that most American Indian tribes are believed to have a high level of diversity (Ward, Alan Redd, Valencia, Frazier, & Paabo, 1993).

As has been discussed, prehistoric migrations are difficult to reconstruct from mtDNA and Y chromosome data. The most meaningful measure of migration from a genetic point of view is obtained by taking the generation as the time unit. Measuring the distribution between birthplaces of parent and offspring theoretically can yield a statistical measure of migration. However, this method works only for a continuous model in which the population is constant, and is not entirely satisfactory when the population is highly clustered as is believed most prehistoric populations were (Cavalli-Sforza & Bodmer, 1971, p. 433). A similar limitation in using such data to infer migrations is that exchange between non-neighboring clusters may have been frequent enough among prehistoric populations to violate the rules of the simplest stepping-stone models (Cavalli-Sforza and Bodmer 1971). See the following chapter for a discussion of migration models.

Finally, it must be mentioned that human mtDNA variation is high, and that genetic variation within populations is much greater than between populations (Walpoff, 1999, p. 551). What this means is that mtDNA evolution, and possibly the evolution of other genetic systems, is not the same as the evolution of particular populations. As Scozzari et al. (1999) have noted, groups or tribes thought to have descended from a common ancestor more than 10,000 years ago may have lost even their shared-by-descent portion of their gene pool and can no longer be determined as biologically affiliated through genetic analysis. Therefore, population specific mutations and the gene trees inferred from these sequences are generally inconsistent with historic and prehistoric population affiliation. Page and Charleston (1990) identified a method for visualizing and quantifying the relationship between a pair of gene and species trees that constructs a third, reconciled tree. Reconciled trees use a more critically optimal method for mapping the combined history of genes and populations. However, even this more accurate method of depicting gene and population trees has limitations, such as allele phylogenies and horizontal transfer, neither of which has been addressed in studies concerning American Indian historic population movements. In fact, many of the polymorphisms observed for mtDNA probably predate population separations (Mountain & Cavalli-Sforza, 1997) and would not be useful in constructing genetic, population, or reconciled trees. It must be remembered that mtDNA or Y chromosome lineages are not human populations. In order to estimate the significance of variation of gene frequencies between groups, it is necessary to estimate how large a sample must be in order to be representative of the group. This can only be accomplished if an accurate estimate of the real variation to be expected in the gene frequencies is possible. Furthermore, this estimation is valid only for genes without dominance, in which case genes can be counted. However, if people in the sample from a given tribal village or town are closely related, a single source of variation may greatly inflate the estimate of variance between populations (Cavalli-Sforza and Bodmer 1971, p. 422). Multivariate analysis, or the use of more than one trait or gene, which is presently the most commonly employed method of analysis, poses more difficult problems in that one must determine the maximum number of genes possible for each population in order to be accurate. Unfortunately, many authors have tested only a small set of markers on one gene (univariate) for their studies (Cavalli-Sforza, Menozzi, & Piazza, 1994, p. 22), combining their data with those of others to arrive at several sets of markers for their multivariate analysis.

CHAPTER SEVEN
HUMAN MIGRATIONS AND MOLECULAR ANTHROPOLOGY

As discussed in the last several chapters, molecular anthropologists have borrowed a large amount of their theoretical methodology from population biology and molecular genetics. This has resulted in several limitations arising within the field of molecular anthropology, primarily in the attempt to make humans, a culturally and biologically based species conform to purely biological models. However, molecular anthropologists have also drawn heavily on physical anthropology and demography, primarily for the development of migration models. This chapter covers several of the key migration models used within the field, focusing especially on those pertaining to the peopling of North America. First, however, a few basic assumptions need to be discussed before the actual models can be covered. First, it must be remembered that a survey of several small societies along a continuum of increasing population density, intensity of land use, and social integration demonstrates that no perfect correlation between position on this continuum and all aspects of migration is apparent. However, some generalizations do emerge from these comparisons.

For instance, population density, particularly at the low end of the scale, strongly conditions the mean marital distance within a society. Where population densities are fewer than one person per square kilometer, such as would have been found in the initial peopling of the Americas, individuals must marry within a much larger area than do spouses from more densely populated places. Clearly this relationship may be affected by the mode of transportation available, but when societies depend on foot, animal, or water transport, a greater population concentration provides a sufficiently large potential spouse pool in a smaller area.

Likewise, the higher travel cost of obtaining spouses from a greater distance may be required of members of low density populations because of the unavailability of potential local spouses. There is good reason to believe, therefore, that a minimum size is necessary in order for a population to be endogamous (MacCluer & Dyke, 1976). Below this minimum, demographic fluctuations in sex ratio, age, and kinship structure reduce the probability of finding a suitable mate. As population density increases, local mate pools become larger and individuals do not need to travel as far to find a spouse.

Similarly, the level of social and cultural integration also influences mobility and migration. Societies in which the family is the highest level of integration are more likely to use mobility as a risk stabilizer. The lack of higher levels of political integration and authority in family-level societies also increases mobility since the ultimate resolution of dispute is to leave the group (Turnbull, 1968; Yellen & Harpending, 1972). With higher densities and more intensive use of land, mobility may be reduced. Thus, when discussing the molecular anthropological data in the investigation of the peopling of North America, these points must be kept in mind. It is reasonable to assume that the initial groups that migrated to the Americas were relatively small in population size. Therefore, most groups probably practiced some form of exogamous spousal exchange. In fact, it may be assumed that this practice was continued up until the time of Euroamerican contact by most American Indian tribes, as evidenced by the ethnographic record. With these points in mind, it is now possible to examine several population genetic models.

Sewall Wright (1931), who is responsible for developing many of the theories discussed below, aimed to produce a mathematical treatment applicable to all species. As a result, the classic population genetic models of migration emphasize generality and mathematical tractability rather than realism

(Levins, 1966). Such generality, as the models have proven, has great advantages, but it also has great costs, especially with humans who possess culture.

Many molecular anthropologists have realized this fundamental limitation, acknowledging that if the deviations from assumed model conditions are not great or the predictions need not be precise, the cost of simplification will be theoretically slight. For example, Morton (1977) claimed that Malecót's (1955) equation relating genetic similarity and distance provided an "acceptable" fit to many human populations. Morton also suggested that this very general model provided "a less complete and reliable prediction" than a more detailed model. Thus, if it is sufficient to know that genetic similarity declines more sharply in continental "isolates" than among hunter-gatherers or oceanic islanders (Morton, 1972), and if geographic distance is the only variable considered relevant, then the model can theoretically summarize the relationship. However, when more careful attention is paid to the levels of subdivision in populations, it can be shown that geographic distance is not the only variable determining genetic similarity and its effect varies with the level of population subdivision (Fix, 1979). Of these theoretically "simple" genetic population models, four are of particular importance in the discussion of the peopling of North America.

The Island Model

As mentioned, Sewell Wright was concerned with developing models applicable to all species, thus the basic population structure underlying much of his mathematical theory is the island model (Wright, 1951; 1969). As a result, many of the predictions concerning the relative importance of genetic drift, natural selection, and mutation are based on inequalities derived from this basic model. Furthermore, crucial formulae specifying the expected degree of inbreeding in subpopulations also rest on the assumptions of the island pattern. For example, consider the commonly cited equation specifying the equilibrium between migration and genetic drift:

$$F = 1/4N_e m + 1$$

Where F is the inbreeding coefficient in the subpopulations, N_e is the effective subpopulation size, and m is the migration rate. From this simple equation comes the prediction that if m is much smaller than $1/4N_e$, then F is large and there is a high degree of local inbreeding and homozygosity, whereas the opposite condition leads to reduced inbreeding and ultimately to panmixis (i.e., random mating) among all subpopulations as a single large breeding population. A commonly cited result of this equation is that a very low migration rate, on the order of one migrant per generation, is needed to counteract local differentiation (Crow & Kimura, 1970; Spieth, 1974). The ramifications of this finding are extensive, but in particular it would imply that genetic micro-differentiation could only occur under conditions of almost complete subpopulation isolation. According to this model, then, only one individual would have to migrate between tribes per generation in order to counteract differentiation between tribes. This, however, was never the case in North America, as the archaeological, linguistic, and ethnographic record attests. For example, along the Northwest Coast and Plateau culture regions, many American Indians practiced a high degree of spousal exchange and intergroup marriage among other groups in order to solidify trade arrangements and political alliances. Some of these exchanges took place well over 500 miles from where the group has been historically recorded to inhabit. Examples of these trade centers are the large Native fisheries of the Northwest Coast such as The Dalles, Celilo Falls, and the Lillooet River Fishery (Hayden, 1992; Schuster, 1998; Stern, 1998) where groups from the Northwest Coast, Plateau, Northern Plains, and Great Basin regions gathered.

Likewise, in the island model, migrants are assumed to have a gene frequency equal to that of the entire assemblage of subpopulations. Note, however, that this frequency is not a random variable and migrants are not a sample of the entire population, therefore this model assumes that the gene frequency of the migrants has no variance, an unlikely situation in the real world. Furthermore, according to this model the gene frequency change in each subpopulation, then, where only migration is considered, will be a function of gene frequency in the subpopulation, p_i, the migration rate, m, (specifically, the proportion of the subpopulation replaced by migrants each generation), and the mean gene frequency of the entire population, p.

That is:
$$Dp = -mp_i + mp$$

The stabilization power of migration is obviously very great in such a system. Each generation local gene frequencies are pushed toward the overall population mean gene frequency by migration. Thus, as m approaches 100%, each subpopulation gene frequency, p_i, will become the population mean frequency, p. Furthermore, m will always decrease local genetic differentiation due to processes such as genetic drift or localized selection.

Theoretically then, this theory would predict that if North America acted as an island, by now American Indians would have the same population genetic frequency because although the various tribal genetic frequencies may have been different during the initial peopling of North America, by now enough migrations between tribes would have taken place to equal out the subpopulation differentiations. The limitations of this theoretical model are quite apparent when discussing the peopling of North America. The ancestors of present-day American Indians were never isolated groups, and according to the ethnographic record, as well as what can be determined from the archaeological and linguistic record, every population group practiced various forms of exogamous spousal exchange.

Isolation by Distance

At the opposite end of the spectrum of population models from the "isolated" island model is the continuously distributed uniform population implied by Wright's isolation by distance (IBD) model (Wright, 1943). There is no analog in the IBD model to the randomly mating "islands" receiving migrants from an infinitely large panmictic source as hypothesized in the island model. In the IBD model, the population is not subdivided into demes that exchange migrants or receive migrants nor is it a panmictic unit itself. Random mating is hypothesized to be limited by spatiotemporal distance such that individuals are more likely to encounter and mate with neighbors than with those farther away. As a result, groups of individuals may thus be clustered into "neighborhoods," areas defined by "central individuals" whose parents may be treated as if drawn at random (Wright, 1969, p. 295). Genetic variation within the population, therefore, will depend on the size of these neighborhoods, which is also to say that the mobility of individuals or the distance from which they choose mates, defines the spatial structure of genetic variation.

The simplest representation of IBD is a linear (one-dimensional) environment such as a major river corridor along the length of which the difference between parents and offspring birthplaces are normally distributed from one generation to the next (Wright, 1969, p. 297). An example from North America would have been the prehistoric tribal distributions found along the St. Lawrence, Mississippi, Colorado, Rio Grande, and Columbia Rivers. In these cases, neighborhood size is a function of the standard deviation of the parent-offspring difference. More complicated mathematically is the two-dimensional case of a uniform area of individuals; conceptually, however, the model is identical. The effective population size in a uniform area is equal to $2^{\wedge}?^2$, where ? is the standard deviation of parent-offspring birthplaces. Thus, the IBD model may work for specific questions in the Plateau, Plains, and parts of the Southwest culture regions prior to the American Indian tribes acquiring the horse in the seventeenth and eighteenth centuries (Haines, 1938a; 1938b). However, once they acquired the horse, the American Indians of these regions were no longer confined to the river corridors and the surrounding environments, and in many cases traveled hundreds of miles.

Wright (1969) also considered the case when the distribution of parent-offspring birthplaces is not normal, especially for the often observed situation where most mates (and resulting offspring) are found at close distances. While perhaps more realistic than the island model, the assumptions of IBD are almost never met in real human populations. Regional populations of hunter-gatherers are so fluid that more likely they approximate a continuous distribution of individuals. Similarly, areas of dense populations, for example intensive farmers or those living along a river corridor who are undivided by linguistic or political boundaries, might also nearly satisfy the assumptions for IBD, but as noted, these conditions have not been apparent in North America except for the far north along the Mackenzie and Yukon Rivers for the past 250+ years or more.

Malecót's Isolation by Distance Model

Malecót's (1955; 1973) treatment of isolation by distance is the continuous population model most widely applied to humans. Cavalli-Sforza (1984) suggests that the popularity of this approach is due to its mathematical simplicity compared to the Wright's approach. Morton (1973; 1977) and his colleagues redefined the concept of "population structure" is the fitting of the Malecot equation to human data, and numerous populations ranging from hunter-gatherers to modern state societies have been studied (Cannings & Cavalli-Sforza, 1973; Cavalli-Sforza, 1984; Felsenstein, 1975; Fix, 1979).

Malecót's model assumes a continuous uniform population with migration simply being a function of geographic distance. Genetic similarity, defined as a coefficient of kinship #, declines with distance following a negative exponential function:

$$\#(d) = ae^{-bd}$$

The parameters of this model, a and b, defined as "local kinship" and "rate of decline" respectively, are estimated by

$$a = 1/(1 + 4N_e ?m^{1/2})$$

where N_e is the effective population size, $?$ is the standard deviation of marital migration distances, and m is called "systematic pressure," and

$$b = (8m)^{1/2}/?$$
despite the fact that these quantities are actually estimated by regression from the data in empirical studies (Jorde, 1980; 1985; Jorde, Bamshad, & Rogers, 1998). Later additions to the model included adjustments for the dimensionality of the environment (although again these were little used in actual applications) and the addition of a correction factor, L, to compensate for the fact that #, a similarity measure, was often negative for long distances (Jorde, 1980).

The basic similarity of this model to Wright's IBD is apparent - both see genetic similarity developing in local neighborhoods as a function of the degree of dispersal. Wright's basic model considers the distribution of parent-offspring birthplace differences to be normal while Malecót's treats marital distances as a negative exponential distribution. Likewise, Wright defined genetic similarity as a correlation while Malecót defined it as a probability or kinship coefficient. For individuals, the kinship coefficient is defined as the probability that two random homologous genes will be identical (Crow & Kimura, 1970, p. 68). The difference is a result of two different definitions of the inbreeding coefficient: one as the correlation of uniting genes; the other the probability of identity by descent (Jacquard, 1975).

Malecót's formulation also shares the emphasis on generality and simplicity that Wright's models stressed. It is an obvious fact of any human population that geographic distance is an important determinant of human movement but, as the legacy of cultural anthropology has demonstrated, it is not the only variable that structures human movement. Malecot's model further reduces this relationship to a linear, one-dimensional case (Cavalli-Sforza, 1984; Cavalli-Sforza, Menozzi, & Piazza, 1994), resulting in a highly unrealistic assumption since most human populations exchange mates with neighbors on all sides. The model also reduces the distribution of marriage distances in all human populations to the negative binomial (see Swedlund, 1980).

More problematic than these simplifying assumptions, however, is the parameter, m, the "systematic pressure."

Although perhaps unrealistic, the exponential distribution of mating distances is at least a clearly stated and easily comprehended assumption. In contrast, m conflates several potential evolutionary processes including mutation, stabilizing selection, and "long-distance" migration. Leaving aside mutation, which as Cavalli-Sforza (1984) notes, may be negligible in its effects on genetic similarity at this scale, and selection, which for particular alleles may have crucial effects on any geographic scale, the concept of long-range migration as a stabilizing force has serious problems. The ability of this form of migration to overcome the random differentiation of local regions follows directly from the assumption of the island model that migrant gene frequencies are an average of the overall population and constant through time (see above). This assumption exists in the absence of any actual data on patterns of long-range human migration or the gene frequencies of such migrants. It is certainly possible to argue that a small number of random migrants from distant locales might represent highly unrepresentative samples of their parent populations. Such migration, therefore, might contribute an overall stochastic rather than stabilizing effect to the population system. Just as the founder effect is seen as a form of random genetic drift, migrant groups may also constitute a random evolutionary process. In any case, if populations are truly distributed continuously with no discontinuities between neighboring groups, then the distinction between local and long-distance dispersion is simply arbitrary (Cavalli-Sforza, 1984).

On a positive note, the application of Malecot's IBD model has provided a very general view of the pattern of human genetic variation and, as Cavalli-Sforza (1984) and Morton (1977) have indicated, the picture is consistent with common sense expectations. Thinly distributed populations such as is believed to have been the case during the peopling of North America show a slow decline of similarity with distance; continental "isolates" show a sharper decline. However, as Cavallis-Sfroza (1984) also states, to describe human population structure using a model of only two parameters is an extreme type of reductionism.

The Stepping-Stone Model

Another model commonly used is the stepping-stone model, which is a metaphor for the pattern of discrete subdivisions connected by migration that characterizes this model (Kimura & Weiss, 1964). In this model, each subpopulation exists either in a one-dimensional array or a two-dimensional lattice and exchanges migrants with its nearest neighbors (two in the linear case; four in the plain case). In contrast to the previously described discrete island model, the stepping-stone model incorporates the IBD criterion seen in the continuous models. The subdivision of the population system into discrete stepping-stones, however, may more closely approximate the colonies or local demes of many natural populations including humans (Crow & Kimura, 1970). Although more realistic in this sense, this model retains the basic simplifying assumptions of the other mathematical models of population structure: infinite number of colonies (corresponding to the infinite extent of the continuous population in IBD models); and, symmetric and isotropic migration of constant intensity persisting to produce genetic equilibrium (see Jorde, 1980). Also shared with other general models is the notion of a systematic stabilizing force that might be mutation and/or selection, but is usually described as long-range migration (Crow & Kimura, 1970). Just as for the island model, each colony theoretically receives a proportion of migrants from the combined gene pool of the entire system. Since this metapopulation is very large, it is theoretically not subject to genetic drift and therefore the migrant gene frequency hypothetically remains constant from generation to generation. This convention makes each colony of the stepping-stone system equivalent to an island and, in the case of no local migration, the stepping-stone model formally reduces to a series of independent islands.

Therefore, the stepping-stone model makes similar predictions to other IBD models regarding the decline of genetic similarity with distance (not surprising since the model may be made equivalent to the continuous IBD model by decreasing the inter-colony distance to zero (Jorde, 1980). As a result, the principal use of the model has been theoretical in contrast to the wide application of the Malecót model to empirical human populations. Furthermore, as with most theoretical applications, predictions concerning the pattern of genetic similarity or the decay of heterozygosity depend on constancy of conditions until equilibrium is reached. Experience with actual human populations, however, makes these assumptions extremely problematic. Population sizes and migration rates often vary through time and are conditioned by location in space (i.e., non-isotopic). Thus, depending on the rapidity of convergence in equilibrium, conditions may change drastically in a human population system.

It should be clear that this model has many limitations when applied to the peopling of North America. It is highly unlikely that at any point in prehistory were American Indian population sizes constant, nor is it likely that migration rates were constant. There are also basic limitations to all of these models that cultural anthropological study has demonstrated. Social and kinship data and historical records are also basic variables in migration models that must be considered, as well as the ranges of variation in these variables among, and between, societies and cultures. Thus one important aspect of migration is the stage of the life cycle at which it occurs: premarital; marital; or post-marital. Genetic models are concerned with intergenerational gene flow and ignore these differences in the timing of the migration.

However, it is essential that these basic variables be included if the migration model is to have some validity. For example, if most dispersal occurs after the high mortality stage of the life cycle, which is the case in the widespread human pattern of marital or post-marital movement, the standard predictions of migration models may not hold. When the dispersing group is small, the migrants constitute a statistical sample of the parental gene pool. As Rogers (1988) has pointed out, under these circumstances, migration is formally equivalent to a component of genetic drift. Rather than homogenizing genetic variation among exchanging populations, such stochastic migration can actually increase gene frequency variance.

Similarly, the migration models discussed above do not specify the units of migration but subsume the scale of migration under the value of m, the migration rate, irrespective of whether individuals or groups do the traveling. These models fail in this sense to consider the potential effects of the scale of migration for the structure of migration. Human populations that periodically fission or split off groups that migrate to fuse with other populations differ from the implicit model of random individuals migrating between demes. As for the life cycle stage of migration, the migration of groups opens the door for stochastic effects on genetic variation.

In particular, migrant groups can be structured along kin lines, such as when families migrate together. Since biological kin are likely to share many genes in common, this pattern of migration can lead to highly biased genetic samples of the donor population gene pool. Depending on the degree of relatedness of group members, kin-structured migration may augment, rather than reduce, differences between donor and recipient populations. Indeed, when stochastic effects are shared, or negatively shared (Epperson, 1994), dramatic effects on spatial patterns of genetic variation can result.

Furthermore, emphasis in the models discussed is on equilibrium patterns of genetic variation achieved after many generations of constant rates and patterns of migration. Population sizes of donor and recipient populations in these models are assumed to remain constant and gene flow must be balanced among groups to maintain this constancy.

These basic, yet simple, factors are often neglected within the models employed by molecular anthropologists when investigating the peopling of North America. The history of human populations records growth and decline through time, changing migratory patterns, and large-scale movements of peoples and trade contacts between distant locales. Gene flow resulting from these long distance contacts may have been crucial in the spread of adaptive alleles such as the hemoglobin variants (Livingstone, 1989). Sustained growth can lead to range expansion involving the movement of colonists into unoccupied territory. This "migratory" process can have numerous genetic effects ranging from founder effect, or the augmented founder effect called "lineal effect" (Neel & Salzano, 1967), to the process Cavalli-Sforza, Menozzi, and Piazza (1993) have dubbed "demic diffusion" where invaders absorb the previous inhabitants of a territory.

Therefore, as has been discussed, the field of molecular anthropology uses a wide array of terms and concepts derived from molecular genetics (only some of which have been discussed here (only some of which have been discussed here, see Jones, 2002 for more on limitations; also see Jones, 2004). This use of unfamiliar terms can result in a misunderstanding of the data, because the underlying assumptions that the data represent are not clearly described. Furthermore, the last few chapters have also attempted to elucidate some of the fundamental principles involved in the use of mtDNA and Y chromosome data and the investigation of American Indian historic population movements. With these points in mind, it is now possible to examine the various studies that have been conducted and what they have found.

CHAPTER EIGHT
AMERICAN INDIAN MTDNA AND
Y CHROMOSOME STUDIES

As previously mentioned, the majority of studies involving American Indian mtDNA and Y chromosome data have been concerned with the initial peopling of the Americas. There have been only a handful of studies that have attempted to study human population movements within the Americas, which will be discussed below under the chapter covering ancient DNA (aDNA). First, however, studies using classical genetic markers and American Indians will be briefly reviewed to establish a baseline upon which mtDNA and Y chromosome data can be contextualized.

Studies have shown that American Indians resemble Siberian and other Asian populations in the kinds of frequencies of various genetic markers of the blood they carry. For a more complete description of classical genetic markers and their use in tracing populational affinities and origins, see Crawford (1973). Szathmary (1993) has summarized the genetic diversity of North American Indian populations based upon classical frequency distributions. American Indian and northeastern Siberian populations have similar frequencies of many blood types, forms of serum proteins, and red-cell enzymes. More recent research has confirmed that mtDNA and Y chromosome haplotype and haplogroup distributions are also similar. When compared to other geographical populations of the world, on the basis of multivariate statistical analyses of gene frequencies, the Siberian or Asian populations tend to cluster together with those of the American Indians.

In 1988, Cavalli-Sforza et al. used an average linkage

analysis of Nei's genetic distances to construct a genetic tree based upon 120 alleles from 42 world populations. A bootstrap method (a resampling technique for obtaining standard errors) was utilized to test the reproducibility of the sequence of the splits in the phylogenetic tree (dendrogram). This tree showed two main branches, the African and non-African. The North Eurasian branch divided into Europeans (Caucasians) and Northeast Asians, including the American Indians. Thus, this multivariate approach to population affinities revealed a close genetic relationship between American Indians and Asian groups.

However, not all classical genetic markers occur across the world in differing frequencies. Instead, some occur only in American and Asian populations. These include the following: the Diego allele, DI*A; gamma globulin allotypes, GM*A T; Factor 13B*3; transferring, TF*C4; and complement, C6*B2 alleles. Szathmary (1993) added SGOT*2 (glutamic oxaloacetic transaminase), TF*D, GC*TK1 (GC 1A9), and GC*N (GC 1A3), to a list of markers that indicate an Asian connection. Although the Diego DI*A gene is not always present in all American Indian groups, when it is detected DI*A occurs only in American Indians or Asians. The frequency of the immunoglobulin haplotype GM*A T in Asian populations reaches 50 percent in central Mongolia but is at a lower frequency in North American Indian groups. Similarly, GM*A G is found at frequencies varying between 86 percent in the Chukchi of Siberia (Schanfield, Crawford, Dossetor, & Gershowitz, 1990) to 56 percent among the Ainu of Japan (Matsumoto & Miyazaki, 1972). In North American Indian populations, this GM marker varies from 98 percent among the Northern Cree to 47 percent in a mixed Alaskan group (Schanfield, Crawford, Dossetor, & Gershowitz, 1990). Less is known about the geographic distribution of the complement B2 allele and Factor 13B*3; however, preliminary analyses suggests that these alleles occur at higher frequencies in both the American Indian and Asian groups.

In many of the other genetic systems, such as the human leukocyte antigen (HLA) system, the various blood groups, and even the mitochondrial DNA (mtDNA) Asian haplotypes, most of the forms occur in some other populations of the world, but at different frequencies. Of this last genetic system, American Indians share the four major haplogroups (A-D) with Asian populations (Torroni & Wallace, 1995). In addition, Siberian and American Indian populations share two identical mitochondrial DNA haplotypes, namely S26 (AM43) and S13 (AM88). The S and AM designations represent the same haplotypes, defined by the presence or absence of the specific restriction sites, in Siberian and American Indian populations. From these two haplotypes, Torroni et al. (1993) attempted to reconstruct the time of divergence of the Asian and American Indian mtDNA variation. These differences in the frequencies of some of the genetic markers led these researchers to conclude that American Indian populations are the result of small founding groups, unique historical events in a sense, and possibly the action of natural selection over a span of 15,000 to possibly 40,000 years.

Within the area of physical anthropology and its use of classical genetic data, William Boyd (1952) was one of the first to compare American Indians to other world populations. He believed that American Indians as a whole were distinct from other major continental populations in their blood-group frequencies. He proposed a single American Indian serological grouping, one of seven such major groupings. In the volume

High-incidence markers	Low-incidence or absent markers
ABO*O	ABO*A2
MN*M	ABO*B
RH*R1	RH*R0
FY*A	LU*A
DI*A	K*K
ABH*SE	LEA*

TABLE 1: Source: After Layrisse (1968)

Biomedical Challenges Presented by the American Indian (1968) Miguel Layrisse summarized the distinctive patterns of frequencies in American Indian populations (see Table 1).

During the past two decades, innovations in biochemical genetics and serology have produced a plethora of genetic markers that can be utilized to evaluate populational affinities and movements. Since Layrisse's compilation of 1968, a number of new genetic markers have been identified through electrophoresis, isoelectric focusing, and immunologic techniques. Two of the most informative sets of genetic markers encode the immunoglobulins (GMs and KMs) and the human leukocyte

High-incidence	Low-incidence or absent
HLA*A2, HLA*A9, HLA*W28, HLA*A10	HLA*A1, HLA*A3, HLA*A11
HLA*BW15, HLA*BW16, HLAA*BW40	HLA*B29, HLA*B18
GM*AG, GM*AT	GM*FB, GM*A,FB
GC*1S, GC*IGLOOLIK	BF*F
GC*CHIP	
ALB*MEX, ALB*NASK	
TF*DCHI, TF*BO-1	
CHE1*S, CHE*2+	
TABLE 2	

antigens (HLA system). The newer markers that distinguish American Indian populations are listed in Table 2.

According to Lampl and Blumberg (1979), the HLA*A2 allele has the highest incidence among American Indians. In addition, HLA*A9, HLA*W28, and HLA*W31 are common alleles. Bodmer and Bodmer (1973) note the absence of HLA*A1, HLA*A3, HLA*A10, HLA*A11, HLA*W29 from American Indian populations. Furthermore, North and South American Indian populations can be distinguished on the basis of HLA*AW31 and HLA*W15, which occur at high frequencies among South American Indian groups, whereas HLA*W28, HLA*A9, and HLA*W5 are North and Central American markers.

Other genetic systems that can be used to establish population affinities between Asian and American groups include: group-specific component (GC), serum pseudocholinesterase (CHE1 and CHE2), properdin factor B (BF), transferring (TF), and albumin (ALB). However, these classical markers are all rather crude when attempting to investigate historic population movements, especially within an area such as North America. DNA RFLPs and sequences of mitochondrial and Y chromosome DNA, on the other hand, are providing finer discrimination among populations and individuals. Furthermore, with recent developments in the use of restriction fragment length polymorphisms (RFLP), amplification through polymerase chain reaction (PCR), and nucleotide sequencing, it is now possible to explore the variation of the DNA molecule in greater detail (Cann, 1985).

Mitochondrial DNA (mtDNA)

As already described, mitochondrial DNA (mtDNA) is a small, circular molecule of 16,569 nucleotide pairs (np) located in mitochondria of the cytoplasm (Anderson et al., 1981). This molecule evolves by the accumulation of mutations in the maternal lineages (Brown, Prager, Wang, & Wilson, 1979), and is believed to fix new mutations more than ten times faster than nuclear DNA (Wallace et al., 1987). This rapid evolutionary change allows molecular anthropologists to use it to study the evolutionary divergence of human populations.

Even though Alan Wallace and Rebecca Cann first applied mtDNA to the study of human phylogeny, Douglas Wallace and his research group were the first to use mtDNA to examine questions concerning the peopling of the Americas. Wallace et al. (1985) utilized a series of restriction endonucleases, HpaI, BamHI, HaeII, MspI, AvaII, and HincII to identify haplotypes of mtDNA. Their initial restriction analysis of the mtDNA of the Pima and Papago Indians revealed a distinctive marker HincII morph-6, in 42 percent of the people tested. This haplotype is observed at low frequency in the mtDNA of Asians and its presence at high frequencies among American Indians can be explained by the founder effect (Schurr et al., 1990). Among molecular anthropologists there is some controversy surrounding the number of maternal lineages that are necessary to account for the observed variation in mtDNA haplotypes of American Indian populations. Initially, Schurr et al. (1990) argued for the presence of four different lineage haplogroups A – D, as follows: 1) HincII morph-6 (site loss at nt 13,259 and an AluI site gain at nt 13,262) marker (haplotype AM10) found in virtually every mtDNA from Pima, Maya, and Ticuna Indians, later termed haplogroup C. 2) Asian-specific COII-tRNA^{Lys} intergenic deletion (haplotype AM2), which is found in American Indians who lack the HincII morph. This is defined by a 9-bp deletion and is referred to as haplogroup B. 3) HaeIII site at nt 663 polymorphism (haplotype AM6) first observed by Cann (1982) in Chicanos and Chinese, but was later noted in the Pima, Maya, and Ticuna. This lineage is now referred to as haplogroup A. 4) Haplogroup D is defined by an AluI site loss at nt 5176. Schurr et al. (1990) placed this haplogroup in the middle of the American Indian phylogenetic tree. Because most of these haplotypes can be derived from each other with the sequential accumulation of mutations, Wallace et al. (1985) initially claimed the origin of all American Indians from a single lineage. Pääbo et al. (1988), after an analysis of the mtDNA from a 7000 year old American Indian brain from Florida, proposed the existence of another founding lineage. However, Schurr et al. (1990) explained these results in terms of the loss of this founding haplotype from the three American Indian tribes they had studied.

A more extensive analysis by PCR amplification with 14 endonucleases of 321 American Indians from 17 populations confirmed the presence of the four different lineages or haplogroups A, B, C, and D that account for 96.9 percent of American Indian mtDNA variation and are of Asian ancestry (Torroni, Schurr et al., 1993; Torroni & Wallace, 1995). These findings supported the hypothesis that the four American Indian mtDNA haplogroups resulted from four separate demic expansions. Three of the four haplogroups (A, C, and D) observed in the Americas are present in indigenous Siberian populations (Torroni, Sukernik, Schurr, Starikorskaya, Cabell, Crawford, Comuzzie, & Wallace, 1993). Presently, none of the Siberian populations have exhibited haplogroup B. The presence of haplogroup B in Asia and the Americas, and its absence from Siberia, is suggestive of its separate expansion into the Americas, possibly prior to the peopling of Siberia (circa 20,000 years ago). When dates of divergence were calculated using the mtDNAs of American Indians and Siberians, they fell between 17,000 and 34,000 ybp (Torroni, Sukernik, Schurr, Starikorskaya, Cabell, Crawford, Comuzzie, & Wallace, 1993).

Torroni and Wallace (1995) reported that out of 743 American Indians tested, 25 individuals, scattered among eight tribal groups of North, Central, and South America, displayed some mtDNA variants that differed from the four common haplogroups (A, B, C, D). They suggested that these variants may be the result of: 1) a second mutational event; 2) possible admixture with Europeans or Africans; or 3) additional Asian haplogroups brought into the Americas by Siberians. Thus, Torroni and Wallace cautioned researchers against classifying mtDNAs from Asian populations by using only the primary variants found in American Indian mtDNA. They went on to point out that the 9-bp deletion had occurred independently in different regions of the world. This conclusion was supported by Soodyall et al. (1996), who discovered the presence of the so-called Asian-specific deletion in sub-Saharan Africa. From these data it appears that this 9-bp deletion arose independently at least twice, once in Asia and once in Africa. Furthermore, Bailliet et al. (1994) have proposed the existence of as many as ten possible mtDNA founder haplotypes in American Indian populations. However, others believe that some of these haplotypes may be due to mutations in American Indian populations and/or admixture with Europeans (Bianchi & Rothhammer, 1995).

Ward and his colleagues (1991) sequenced a 360-nucleotide segment of the mtDNA control region from 63 individuals of the Nuu-Chah-Nulth (or Nootka) from Vancouver Island. They identified 28 mtDNA lineages as defined by 26 variable positions within the control region. Ward et al. (1991) computed the average sequence divergence between the lineage clusters using a maximum rate of evolution of 33 percent divergence per million years for the control region. They obtained a range of 41,000-78,000 years, with an average of 60,000 years. These data suggest that the mitochondrial lineages within a single American Indian tribe diverged approximately 60,000 years ago. They interpreted these data as evidence that the lineages were established prior to the American Indian entry into the Americas, and they concluded that the founding populations of American Indians contained considerable genetic diversity.

Merriwether et al. (1995) extensively investigated the geographic distribution of the four founding mtDNA lineage haplogroups in American Indian populations. They observed a north-south increase in the frequency of haplogroup B, accompanied by a north-south decrease in the frequency of haplogroup A. Based upon the extensive distribution of the four lineages in the Americas, Merriwether and his colleagues concluded that the pattern is consistent with a single migratory wave from Siberia into the Americas, followed by genetic divergence. However, these data can also be interpreted to represent a number of migrations from Siberia reintroducing the same haplogroups.

After this study, Merriwether et al. (1996) went on to compare mtDNA RFLPs from Mongolians of Ulan Bator with an array of frequencies of the founding lineage haplogroups in American, Asian, and Siberian populations, revealing considerable similarity between the Mongolian and American populations (Merriwether, Hell, Vahlne, & Ferrell, 1996). In this study the haplogroups were further subtyped into A1, A2, B1,

B2, C1, C2, D1, D2, X6, X7, and "others." Unlike the Northeastern Siberian populations, this Mongolian sample exhibited all four of the American primary haplogroups and shared the highest number of haplotypes with American Indian populations. As mentioned, haplogroup B has not been detected in any of the Siberian populations in closest proximity to the Bering Strait. However, this haplogroup occurs at a frequency of 75 percent among the Atacameno and 50 percent among the Pima, but is absent in a number of other American Indian groups (e.g., Makiritare, Dogrib, and Haida).

The vast majority of mtDNAs from modern American Indian populations belong to primarily five different haplogroups, which have been designated A–D and X (Brown, Hosseini, Torroni, Bandelt, Allen, Schurr, Scozzari, Cruciani, & Wallace, 1998; Forster, Harding, Torroni, & Bandelt, 1996; Schurr, Ballinger, Gan, Hodge, Weiss, & Wallace, 1990; Torroni, 1994; Torroni, Chen et al., 1993; Torroni et al., 1994; Torroni, Schurr, Cabell, Brown, Neel, Larsen, Smith, Vullo, & Wallace, 1993; Torroni, Schurr, Yang, Szathmary, Williams, Schanfield, Troup, Knowler, Lawrence, Weiss et al., 1992; Torroni, Sukernik, Schurr, Starikorskaya, Cabell, Crawford, Comuzzie, & Wallace, 1993). Each of these is distinguished by a unique combination of coding region RFLPs and HVR-I sequence polymorphisms. Together, they comprise 95–100 percent of all mtDNAs in indigenous populations of the Americas (Schurr & Sherry, 2004; Schurr & Wallace, 2002). The same pattern of variation is also observed in ancient Amerindian samples (Carlyle, Parr, Hayes, & O'Rourke, 2000; Fox, 1996; Kaestle, 1995; 1997; 1998; Kaestle & Smith, 2001a; Lalueza, PerezPerez, Prats, Cornudella, & Turbon, 1997; Merriwether, Rothhammer, & Ferrell, 1994; Monsalve, Edin, & Devine, 1998; O'Rourke, Hayes, & Carlyle, 2000b; Parr, Carlyle, & ORourke, 1996; Ribeiro-dos-Santos, Guerreiro, Santos, & Zago, 2001; Ribeiro-dos-Santos, Santos, Machado, Guapindaia, & Zago, 1997; Stone & Stoneking, 1998). Therefore, these five haplogroups are clearly the main founding mtDNA lineages in American Indian populations. However, a certain number of haplotypes not belonging to these five maternal lineages have been detected in different

American Indian groups (Bailliet, Rothhammer, Carnese, Bravi, & Bianchi, 1994; Easton, Merriwether, Crews, & Ferrell, 1996; Lorenz & Smith, 1996; Lorenz & Smith, 1997; Merriwether, Rothhammer, & Ferrell, 1995; Merriwether, Rothhammer, & Ferrell, 1994; Ribeiro-dos-Santos, Guerreiro, Santos, & Zago, 2001; Rickards, Martinez-Labarga, Lum, De Stefano, & Cann, 1999). Most of these have been shown to belong to either haplogroup X, derive from haplogroups A–D mtDNAs, or result from non-native admixture (Schurr & Wallace, 1999; Schurr & Sherry, 2004; Schurr & Wallace, 2002; Smith, Malhi, Eshleman, Lorenz, & Kaestle, 1999). The remaining haplotypes have not been sufficiently analyzed to determine their haplogroup status (Bailliet, Rothhammer, Carnese, Bravi, & Bianchi, 1994). Haplogroup A–D mtDNAs are observed in indigenous populations from North, Central, and South America. Haplogroup A–D mtDNAs have also been detected in populations representing the three American Indian linguistic groups (Amerind, Na-Dene, Eskimo-Aleut) proposed by Greenberg (1987). Among Amerindians, haplogroup A generally occurs at higher frequencies in North America relative to other regions, whereas haplogroups C and D generally occur at higher frequencies in South America. There does not appear to be a distinct clinal distribution for haplogroup B, but it is virtually absent from northern North America (Fox, 1996; Malhi, Schultz, & Smith, 2001; Malhi & Smith, 2002; Schurr, Ballinger, Gan, Hodge, Weiss, & Wallace, 1990; Torroni, 1994; Torroni, Chen, Scott, Semino, Santachiarabenerecetti, Lott, & Wallace, 1993; Torroni, Chen, Semino, Santachiara-Beneceretti, Scott, Lott, Winter, & Wallace, 1994; Torroni, Schurr, Cabell, Brown, Neel, Larsen, Smith, Vullo, & Wallace, 1993; Torroni, Schurr, Yang, Szathmary, Williams, Schanfield, Troup, Knowler, Lawrence, Weiss, & et al., 1992). In contrast, haplogroup X is found nearly exclusively in North America (Brown, Hosseini, Torroni, Bandelt, Allen, Schurr, Scozzari, Cruciani, & Wallace, 1998; Malhi, Schultz, & Smith, 2001; Malhi & Smith, 2002; Torroni, Schurr, Cabell, Brown, Neel, Larsen, Smith, Vullo, & Wallace, 1993; Torroni, Schurr, Yang, Szathmary, Williams, Schanfield, Troup, Knowler, Lawrence, Weiss, & et al., 1992). These distributions likely reflect both the original pattern of settlement of the Americas and the subsequent genetic differentiation of American Indian populations within these continental areas. Although haplogroups A–D usually appear together in Amerindian populations, many tribes lack haplotypes from at least one of these haplogroups (Batista, Kolman, & Bermingham, 1995; Lorenz & Smith, 1996; Lorenz & Smith, 1997; Scozzari et al., 1997; Torroni, 1994; Torroni, Chen, Scott, Semino, Santachiarabenerecetti, Lott, & Wallace, 1993; Torroni, Chen, Semino, Santachiara-Beneceretti, Scott, Lott, Winter, & Wallace, 1994; Torroni, Schurr, Cabell, Brown, Neel, Larsen, Smith, Vullo, & Wallace, 1993). This pattern likely reflects the extent to which genetic drift and founder events have influenced the distribution of mtDNA haplotypes in American populations. However, ancestral populations for the Na-Dene Indians and Eskimo-Aleuts may not have possessed all four of these haplogroups. These populations show different haplogroup profiles than Amerindians, which consist largely of haplogroup A and D mtDNAs. In addition, they essentially lack haplogroup B and have very low frequencies of haplogroup C. Moreover, none of them have haplogroup X (Rubicz, Schurr, Babb, & Crawford, 2003; Saillard, Forster, Lynnerup, Bandelt, & Norby, 2000; Starikovskaya, Sukernik, Schurr, Kogelnik, & Wallace, 1998; Ward, Alan Redd, Valencia, Frazier, & Paabo, 1993). Thus, circumarctic groups appear to have experienced different population histories than Amerindians.

Y Chromosome

The non-recombining (Y-specific) portion of the human Y chromosome has been of great interest to molecular anthropology in reconstructing human phylogeny. Much like mtDNA, but a male mirror image, the Y-specific portion also evolves through the accumulation of mutations, and therefore, markers on the Y-specific region provide some indication of male migration and admixture. The initial research using this data was somewhat disappointing because of the paucity of variation in the Y chromosome. More recently, however, Y-specific polymorphisms have successfully been used to construct informative haplotypes that are specific to geographic regions and to possible historic population movements. In addition, Y-chromosome-specific deletions and transitions have been discovered that apparently have arisen once in human evolution and serve as markers for phylogenetic reconstruction (Karafet, Zegura, Vuturo-Brady, Posukh, Osipova, Wiebe, Romero, Long, Harihara, Jin, Dashnyam, Gerelsaikhan, Keiichi, & Hammer, 1997).

Underhill et al. (1996) reported a C to T point mutation at the DYS19 microsatellite locus. To date this mutation has been found only in Inuits and Navajos of North America and other populations of South and Central America. This mutation is believed to have occurred in Siberia and brought to the Americas by the first Asian migrants. Given an average mutation rate of 1.5×10^{-4} and an average generation time of 27 years, Underhill and colleagues (1996) computed the age of this transition at 30,000 years ago. However, they acknowledged that using mutation rates of a smaller magnitude can provide dates that are more recent. This polymorphism, in the future, may shed some light on the colonization of the Americas, particularly if it is present in some Siberian populations and not in others (Santos et al., 1995; Santos, Rodriguez-Delfin, Pena, Moore, & Weiss, 1996).

Lin et al. (1994) investigated the variation in Asian, European, and African-American populations for Y- and X-associated polymorphisms by using the 47z (DXYS5) probe. Although both the X1 and X2 alleles were detected in most of the populations, Y1 and Y2 were polymorphic in only the Japanese, Koreans, and the Hakas and Folo of Taiwan. To date, these X- and Y-associated markers have not been tested in American Indian populations.

In characterizing Y chromosome variation in American Indians, researchers have employed a number of different single nucleotide (SNP) and short tandem repeat (STR) loci to define the paternal lineages present within them (Hammer, 1997; Karafet, de Knijff, & Wood, 1998; Karafet, Zegura, Vuturo-Brady, Posukh, Osipova, Wiebe, Romero, Long, Harihara, Jin, Dashnyam, Gerelsaikhan, Keiichi, & Hammer, 1997; Karafet, Zegura, Posukh, Osipova, Bergen,

Long, Goldman, Klitz, Harihara, de Knijff, Wiebe, Griffiths, Templeton, & Hammer, 1999; Pena et al., 1995; Underhill, Jin, Zemans, Oefner, & Cavalli-Sforza, 1996; Underhill et al., 2001; Underhill et al., 2000). However, these research groups have not used the same combination of genetic markers in their studies, leading to alternative and sometimes confusing nomenclatures for Y chromosome haplotypes and haplogroups. A recent synthesis of these data has resulted in a consensus nomenclature based on known single nucleotide polymorphisms (SNPs) (Consortium, 2002). This system identifies a Y chromosome haplogroup by a letter and the SNP that defines it (e.g., G-M201). American Indian Y chromosome haplotypes derive from subsamples of the haplogroups present in Siberia. These include haplogroups Q-M3, R1a1-M17, P-M45, F-M89, and C-M130. Two of them, Q-M3 and P-M45, represent the majority of American Indian Y chromosomes. Q-M3 haplotypes appear at significant frequencies in most American Indian populations and are distributed in an increasing north-to-south cline within the Americas (Bianchi, Bailliet, Bravi, Carnese, Rothhammer, Martinez-Marignac, & Pena, 1997; Bianchi, Catanesi, Bailliet, Martinez-Marignac, Bravi, Vidal-Rioja, Herrera, & Lopez-Camelo, 1998; Lell, 2000; Lell et al., 1997; Lell et al., 2002; Santos et al., 1999b; Underhill, Jin, Zemans, Oefner, & Cavalli-Sforza, 1996). The STR data from Q-M3 haplotypes also reveal significant differences in haplotype distributions between North/Central and South American populations, suggesting different population histories in the two major continental regions (Bianchi, Bailliet, Bravi, Carnese, Rothhammer, Martinez-Marignac, & Pena, 1997; Bianchi, Catanesi, Bailliet, Martinez-Marignac, Bravi, Vidal-Rioja, Herrera, & Lopez-Camelo, 1998; Lell, 2000; Lell, Brown, Schurr, Sukernik, Starikovskaya, Torroni, Moore, Troup, & Wallace, 1997; Lell, Sukernik, Starikovskaya, Su, Jin, Schurr, Underhill, & Wallace, 2002; Santos, Pandya, Tyler-Smith, Pena, Schanfield, Leonard, Osipova, Crawford, & Mitchell, 1999b; Underhill, Jin, Zemans, Oefner, & Cavalli-Sforza, 1996). Haplogroup P-M45 is also widely distributed among American Indian populations and represents 30 percent of their Y chromosome haplotypes (Bortolini et al., 2002; Bortolini et al., 2003; Lell, Sukernik, Starikovskaya, Su, Jin,

Schurr, Underhill, & Wallace, 2002; Ruiz-Linares et al., 1999). In addition, phylogenetic analysis has revealed two distinct sets of P-M45 haplotypes in American Indian populations. The first of these (M45a) is more broadly distributed in populations from North, Central, and South America, whereas the second (M45b) appears in only North and Central American groups (Lell, Sukernik, Starikovskaya, Su, Jin, Schurr, Underhill, & Wallace, 2002). Most of the remaining Y chromosome haplotypes belong to one of several different haplogroups, and comprise only 5 percent of American Indian Y chromosomes. In general, these haplotypes have limited distributions in the Americas. For example, C-M130 haplotypes have only been detected in the Na-Dene-speaking Tanana, Navajo, and Chipewayan, and the Amerindian Cheyenne (Bergen et al., 1999; Bortolini, Salzano, Bau, Layrisse, Petzl-Erler, Tsuneto, Hill, Hurtado, Castro-De-Guerra, Bedoya, & Ruiz-Linares, 2002; Bortolini, Salzano, Thomas, Stuart, Nasanen, Bau, Hutz, Layrisse, Petzl-Erler, Tsuneto, Hill, Hurtado, Castro-de-Guerra, Torres, Groot, Michalski, Nymadawa, Bedoya, Bradman, Labuda, & Ruiz-Linares, 2003; Karafet, Zegura, Posukh, Osipova, Bergen, Long, Goldman, Klitz, Harihara, de Knijff, Wiebe, Griffiths, Templeton, & Hammer, 1999). In addition, R1a1-M17 haplotypes have only been observed in the Guaymi (Ngobe), a Chibchan-speaking tribe from Costa Rica (Lell, Sukernik, Starikovskaya, Su, Jin, Schurr, Underhill, & Wallace, 2002). Neither of these haplogroups has been detected in South American Indian populations.

Ancient DNA

Although mtDNA and Y chromosome data have proven the most useful for investigating the peopling of the Americas, data generated from contemporary sources proves to not offer a very detailed picture of historic population movements within the Americas themselves. For this reason, ancient DNA (aDNA) is preferred when attempting to reconstruct historic population movements within a specific geographic area across time. The genetic frequencies of the aDNA are compared to the genetic frequencies of contemporary American Indians to see if the frequencies match, and if so, one can reasonably establish biological affiliation; if not, a historic population movement may possibly be inferred.

That DNA in ancient specimens could be extracted and characterized was first demonstrated in nonhuman material in 1984 by Higuchi and colleagues, who identified nucleic acids from a museum specimen of the extinct quagga and showed its phylogenetic affinity to the modern zebra (Higuchi, Bowman, Freiberger, Ryder, & Wilson, 1984). A year later, Paabo (Paabo, 1985a; 1985b) obtained DNA sequence data from a 2400-year-old Egyptian mummy. This result was surprising not only for its demonstration of the remarkable antiquity for which molecular genetic analysis was apparently possible, but also for the large DNA fragment sequenced (>3 kb). Both of these early efforts relied on extracting ancient (a)DNA fragments, cloning fragments into a vector, and then subsequently sequencing the cloned fragments. Following the nearly simultaneous development of the polymerase chain reaction [PCR (a molecular technique that uses the complementary nature of DNA bases and an enzyme involved in DNA replication to produce millions of copies of a single, specific DNA target sequence)] (Mullis & Faloona, 1987; Saiki, Gelfand, Stoffel, Scharf, & Higuchi, 1988), a number of researchers began extracting and characterizing aDNA from geographically dispersed human samples.

Most ancient population samples are composed of several individuals separated by varying periods of time in a restricted geographic area, and therefore they do not conform to standard definitions of a population. If the samples come from a geographically and temporally restricted prehistoric horizon, however, and are associated with a uniform material culture, researchers have treated them as representing multiple, related, continuous lineages. It should be recognized at the outset that this is not properly a population in the traditional sense, and assumptions of standard population or genetic analyses are compromised by such sample composition. It also means that reliable temporal provenience is essential for such samples. With the exception of the Fremont samples from the Eastern Great Basin (Parr, Carlyle, & ORourke, 1996), dating of samples for aDNA research has been neither widely nor uniformly practiced.

An additional problem with aDNA research is less than uniform success in obtaining marker typings on all samples. For example, when using discrete marker data, such as those used to infer Amerindian haplogroups, not all primer sets are likely to be successful on every sample. This complicates the computation of haplogroup frequencies and results in haplogroup and marker frequencies that are discordant.

Stone & Stoneking (1993; 1998) obtained DNA from skeletons of the relatively recent Oneota archaeological complex of western Illinois. Mitochondrial DNA haplogroup diversity in the Oneota samples indicated 31 percent were associated with haplogroup A, 12 percent with haplogroup B, 42.6 percent with haplogroup C, and 8.3 percent with haplogroup D. Six specimens (5.5 percent) were inconsistent with any of the American Indian haplogroups. Two of these were subsequently determined to be of exogenous origin, whereas the remainder represented a fifth founding haplogroup. Of the samples, 52 were sequenced for the HVRI region and found to have a high proportion of singleton mtDNA types (73.9 percent). This is higher than typically observed in modern American Indian populations. It may reflect loss of rare lineages due to drift in small populations (perhaps as a result of population declines at contact), or it may be a characteristic of ancient samples in general, due to sampling of lineages through time (Stone & Stoneking, 1998). Insufficient sequence data on other ancient populations are available to distinguish between these alternatives.

Kaestle (1997, 1998) characterized a series of skeletal samples (~300 - 6000 ybp) from Pyramid Lake and Stillwater Marsh in the Western Great Basin. These samples were genetically indistinguishable based on mtDNA haplogroup analysis. They also proved to be genetically similar to modern Paiute/Shoshone and California Penutian samples, with low-to-moderate frequencies of haplogroups A and B, low frequency of haplogroup C, and high frequency of haplogroup D.

O'Rourke and colleagues assayed mtDNA variation in the Northern Fremont of Utah (Parr, Carlyle, & ORourke, 1996) and prehistoric pueblo peoples of the four corners region in the US southwest (Carlyle, 2000). Of 43 Fremont samples, 40 were directly dated, whereas 8 of 40 ancient Puebloan specimens have been directly dated so far, with both sets of samples dating to approximately 1000 - 2000 ybp. The latter samples are distributed over a larger geographic area and a slightly longer time frame than are the Fremont materials. Nevertheless, the haplogroup profiles of these two geographically proximal ancient samples are similar.

Both are characterized by low to absent frequencies of haplogroup A, moderate-to-high (>50 percent) frequencies of haplogroup B, and low (<15 percent) frequencies of haplogroups C and D. Both the ancient Pueblo and Fremont are also characterized by a few samples that do not conform to the traditional four founding haplogroups and are presumed to represent haplogroup X (Smith, Malhi, Eshleman, Lorenz, & Kaestle, 1999), or an as-yet-undetected contaminant.

Modern North American Indian mtDNA variation is strongly geographically patterned (Lorenz & Smith, 1996), and ancient samples studied to date appear to exhibit the same geographic structure (O'Rourke, Hayes, & Carlyle, 2000b). Thus, the Oneota (Stone & Stoneking, 1993; Stone & Stoneking, 1998) are most similar to modern populations currently inhabiting the central plains and eastern woodlands of North America, as well as an archaeologically recovered Fort Ancient sample from West Virginia (Merriwether et al., 1997; Merriwether, Rothhammer, & Ferrell, 1994). The Western Great Basin samples (Kaestle, 1997; 1998) share greatest similarities to modern populations in Northern California and the northwest Great Basin, whereas the Fremont and ancient Pueblo share mtDNA haplogroup profiles in common with modern Southwestern populations. Thus, aDNA analyses confirm that the observed geographic structure of modern North American mtDNA variation has been temporally stable (>2000 years, and possibly going back into the far past circa 6000+ ybp), and apparently little affected by the dramatic disruptions attendant to contact (O'Rourke, Hayes, & Carlyle, 2000a). The observed geographic and temporal stability of mtDNA discrete markers needs to be confirmed with a greater number of ancient samples and hypervariable-region sequence data, however, before aDNA evidence can be taken with a high degree of confidence.

Fewer ancient samples have been molecularly characterized in Central and South America, but among those that have been studied, the geographic and temporal structure noted in North America appears to be lacking. Merriwether and colleagues (1994, 1997) examined mtDNA haplogroup diversity using discrete marker data in two ancient samples from Northern Chile (Chinchorro and Inca) and the Copan Maya of Honduras. The deletion marker was absent in both Chilean samples, although it is found at high frequencies in the modern populations of the region. The Copan Maya skeletal samples were uniformly haplogroups C or D, whereas the modern Mayan populations of the region are characterized by high frequencies of haplogroups A and B. However, partial typing of some specimens indicates that additional haplogroups are present in the Copan skeletal series (Merriwether, Huston, Iyengar, Hamman, Norris, Shetterly, Kamboh, & Ferrell, 1997). These results are consistent with earlier observations of geographic structure of genetic variation in North, but not South or Central, America, based on classical markers (O'Rourke, Mobarry, & Suarez, 1992).

Merriwether and colleagues (Merriwether & Ferrell, 1996; Merriwether, Rothhammer, & Ferrell, 1995) have argued that the ubiquity of American Indian haplogroups in antiquity argues against multiple migrations of American Indian founders to the Americas. In contrast, it has been suggested (Lalueza, PerezPerez, Prats, Cornudella, & Turbon, 1997) that the absence of mtDNA haplogroups A and B in dental samples of extinct populations of southern Patagonia/Tierra del Fuego indicates separate founding events for different haplogroups. However, only two of the 60 samples are of any appreciable antiquity (4000 – 5000 ybp), the remainder dating to the past two centuries. Lalueza and colleagues note that the nineteenth century in southern South America has been called the "extinction period" (Lalueza, PerezPerez, Prats, Cornudella, & Turbon, 1997; Lalueza-Fox, Calderon, Calafell, Morera, & Bertranpetit, 2001; Lalueza-Fox, Gilbert, Martinez-Fuentes, Calafell, & Bertranpetit, 2003). It is not obvious that samples obtained from populations undergoing decimation and extinction would be representative of precontact groups. Indeed, reduced population size during this period would be expected to be accompanied by reduced genetic variability. In contrast, haplogroup B (as well as lineages A and C) is present in a small series of artificial mummies from Columbia (Monsalve, Groot de Restrepo, Espinel, Correal, & Devine, 1994), whereas HVRI sequence data indicates a diversity of haplogroups in 18 Amazonian skeletons dated between 500 and 4000 ybp (Ribiero-Dos-Santos, Santos, Machado, Guapindaia, & Zago, 1996). In addition to all four of the primary founding Amerindian haplogroups,

Ribiero-Dos-Santos et al (1996) found a heterogeneous group of sequences that appeared related to haplotypes observed in modern American Indians and Asians. These researchers suggest that this indicates substantially greater mitochondrial lineage diversity in American Indians prior to the effects of European contacts (O'Rourke, Hayes, & Carlyle, 2000b; Ribiero-Dos-Santos, Santos, Machado, Guapindaia, & Zago, 1996).

CHAPTER NINE
SUMMARY AND CONCLUSION

While there is little controversy about the number and type of founding haplogroups in the Americas, the ages of these maternal lineages continues to be contested. Early studies of RFLP variation in American Indian populations produced time depths for haplogroups A, C, D, and X between 35,000 – 20,000 ybp (Torroni, 1994; Torroni, Chen, Scott, Semino, Santachiarabenerecetti, Lott, & Wallace, 1993; Torroni, Chen, Semino, Santachiara-Beneceretti, Scott, Lott, Winter, & Wallace, 1994; Torroni, Schurr, Cabell, Brown, Neel, Larsen, Smith, Vullo, & Wallace, 1993; Torroni & Wallace, 1995). These estimates were viewed as reflecting the genetic diversity that had accumulated in the American branches of these mtDNA lineages, hence, the time at which modern humans first entered the Americas. Additional support for these findings came from the fact that American Indian and Siberian populations did not share any specific haplotypes (Schurr & Wallace, 1999; Starikovskaya, Sukernik, Schurr, Kogelnik, & Wallace, 1998; Torroni, Schurr, Cabell, Brown, Neel, Larsen, Smith, Vullo, & Wallace, 1993). By contrast, the age for haplogroup B in the Americas was estimated at 17,000 – 13,000 ybp, suggesting that it was brought to the Americas in a later and separate migration from the earlier one that brought the other four lineages. The age of haplogroup X based on RFLP haplotype data was identical to that of haplogroup B, although it increased in age when estimated from HVS-I sequence data (Brown, Hosseini, Torroni, Bandelt, Allen, Schurr, Scozzari, Cruciani, & Wallace, 1998). Subsequent analyses of HVS-I sequence variation in American Indians provided somewhat different perspectives on the antiquity of these haplogroups. Several of them showed that haplogroups A, B, and C had roughly the same extent of genetic diversification in North America, and that haplogroup B could possibly have been present in the Americas by 30,000 – 20,000 ybp (Bonatto & Salzano, 1997a; Bonatto & Salzano, 1997b; Lorenz & Smith, 1997). The older date also implied that haplogroup B arrived in the Americas around the same time that haplogroups A, C, D, and X did. In fact, most HVS-I studies have provided ages for the four major founding haplogroups that range between 35,000 – 15,000 ybp, with the most recent dates being 14,000 – 12,000 ybp (Shields, 1996; Shields et al., 1993). A recent analysis of coding region sequence variation in American Indian populations has also generated average dates for haplogroups A – D of between 20,000 – 15,000 ybp (Silva et al., 2002). Thus, most molecular studies favor an entry time for these mtDNA lineages that is somewhat earlier than the dates associated with the Clovis lithic culture in North America.

Two other important issues about haplogroup ages have arisen in this debate. The first centers on the question whether the older haplogroup age estimates actually reveal the timing of human expansion(s) into the Americas. As previously discussed, because the genetic divergence or coalescence times of genetic lineages do not necessarily correspond to the timing of population splits, it has been suggested that the older dates may reflect the emergence of these mtDNA lineages in Asia rather than their entry into the Americas (Jones, 2002; Shields, Schmiechen, Frazier, Redd, Voevoda, Reed, & Ward, 1993). On the other hand, only the founding RFLP haplotypes for haplogroup A–D have been shown to be present in both Siberia and the Americas (Schurr & Wallace, 1999; Torroni, Schurr, Cabell, Brown, Neel, Larsen, Smith, Vullo, & Wallace, 1993). These data suggested that the temporal split between the ancestral American Indian population and its Asian precursor mirrored the split in the branches of each respective haplogroup. The second issue is the number of founding haplotypes that were brought with each founding haplogroup. The number of founders present in a genetic lineage will affect estimates of its age because a certain amount of the diversity present in that lineage will have accumulated from each founding type. If more than one founding haplotype was present within a haplogroup, then the age of this mtDNA lineage would be inflated due to the fact that the diversity of haplotypes that accumulated from each founding haplotype was not taken into account in the coalescence estimates. For American Indians, the presence of multiple founding haplotypes would imply that the ages of haplogroups A – D would be less than 30,000 – 20,000 ybp; hence, the colonization date of the Americas would be more consistent with a late entry migration model. As noted above, there appears to be only one founding RFLP haplotype each for haplogroups A–D and X (Brown, Hosseini, Torroni, Bandelt, Allen, Schurr, Scozzari, Cruciani, & Wallace, 1998; Schurr & Wallace, 1999; Torroni et al., 1996; Torroni, Schurr, Cabell, Brown, Neel, Larsen, Smith, Vullo, & Wallace, 1993). These founder haplotypes are the most widely distributed mtDNAs in the Americas, and central to the diversification of their respective haplogroups. However, other investigators have suggested that more than one founding haplotype from haplogroups A – D were among the original set of founding American Indian mtDNAs (Alves-Silva et al., 2000; Bailliet, Rothhammer, Carnese, Bravi, & Bianchi, 1994; Merriwether & Ferrell, 1996; Merriwether, Rothhammer, & Ferrell, 1995; Merriwether, Rothhammer, & Ferrell, 1994; Santos, Hutz, Coimbra, Santos, Salzano, & Peena, 1995; Santos et al., 1999a). Unfortunately, none of these studies have provided additional RFLP or HVR-sequence data to demonstrate that these are actually the same founding haplotypes defined in other studies (Schurr, Ballinger, Gan, Hodge, Weiss, & Wallace, 1990; Schurr & Wallace, 1999; Schurr & Sherry, 2004; Schurr & Wallace, 2002). At the same time, Malhi et al. (2002; 2001; 2002) argue that there could possibly be more than one founder HVS-I sequence for at least some of these haplogroups, due to these sequences being identical to ones present in Asian and Siberian populations. However, they also recognize the difficulty in delineating ancestral sequences from derivative forms that have lost or gained key polymorphisms that distinguish American from Asian sequence motifs, due to the recurrent mutations that typically occur in the mtDNA control region. Thus, additional coding region data must be obtained from these mtDNAs to confirm their status as founder haplotypes.

The methods for dating Y chromosome haplogroups have employed both the SNPs that define them and the STR loci that occur on each Y chromosome. Because SNPs are rare, if not unique, evolutionary events, it is difficult to estimate when they evolved in a particular paternal lineage using only this kind of data. To get around this problem, Underhill et al. (2001) used an average mutation rate estimated from SNP variation in three Y chromosome genes to date the various branches (haplogroups) of their phylogeny.

This estimate of $1.24 \sim 10^{-9}$ produced an age for the major expansion of modern humans out of Africa of ~59,000 ybp. Using this date for the Most Recent Common Ancestor (MRCA) of their SNP phylogeny, Underhill et al. (2000) estimated an average SNP evolution rate of one per every 6,900 years. With this rate, it is possible to tentatively date the origins of the major branches of the phylogeny, as well as other points of SNP haplotype diversification. An alternative strategy for dating the ages of Y chromosome haplogroups is to analyze variation in the faster evolving STR loci that co-occur on each SNP haplotype. In this case, the extent of allelic diversity of a set of STR loci are measured and averaged over all loci, with the average then being multiplied by an STR mutation rate to determine the actual age of the Y chromosome lineage. Recent mutation rates have been estimated across multiple generations of males (meiotic transmissions) in human families. Although these rates vary somewhat depending on the type of STR used for the estimates (di-, tri-, or tetra-), most studies have found that the average mutation rate of Y chromosome STRs is around $2.80 \sim 10^{-3}$ per generation (Kayser, Brauer et al., 2001; Kayser, Krawczak et al., 2001). Considerable effort has been made to estimate the age of Q-M3 haplotypes, given that they appear to signal the initial entry of ancestral populations into the Americas. Using the SNP mutation rate of Underhill et al. (2000), one obtains an age for haplogroup Q-M3 of ~13,800 ybp (Schurr & Sherry, 2004). The estimates made with STR mutation rates have ranged from 30,000–7,600 ybp (Bianchi, Bailliet, Bravi, Carnese, Rothhammer, Martinez-Marignac, &

Pena, 1997; Hammer, 1998; Karafet, Zegura, Posukh, Osipova, Bergen, Long, Goldman, Klitz, Harihara, de Knijff, Wiebe, Griffiths, Templeton, & Hammer, 1999; Underhill, Jin, Zemans, Oefner, & Cavalli-Sforza, 1996). Together, these analyses of Y chromosome variation in American Indian populations do not clearly point to an early or late entry of the Q-M3 lineage into the Americas, but tend to favor the latter. The P-M45 lineage is considerably older than the Q-M3 lineage, which derives from it. Using the SNP mutation rate of Underhill et al. (2000), P-M45 haplotypes are estimated to be at least 30,000 years old. This degree of antiquity is also reflected by their widespread distribution in Siberia and Eurasia (Lell, Sukernik, Starikovskaya, Su, Jin, Schurr, Underhill, & Wallace, 2002; Underhill, Shen, Lin, Jin, Passarino, Yang, Kauffman, Bonne-Tamir, Bertranpetit, Francalacci, Ibrahim, Jenkins, Kidd, Mehdi, Seielstad, Wells, Piazza, Davis, Feldman, Cavalli-Sforza, & Oefner, 2000). In addition, the P-M45 lineage has been present in Siberia long enough to diversify into different subhaplogroups. This is shown by the presence of two different types of Y chromosome haplotypes in American Indians, a central Siberian set (P-M45a) that is shared with all American Indian populations, and an eastern Siberian set (P-M45b) that appears only in American Indians from North and Central America (Lell, Sukernik, Starikovskaya, Su, Jin, Schurr, Underhill, & Wallace, 2002). The ages of several other Y chromosome lineages present in Siberia and the Americas have also been estimated. One of the older lineages in Siberia, K-M9, has been dated at >50,000 ybp (Karafet, Zegura, Posukh, Osipova, Bergen, Long, Goldman, Klitz, Harihara, de Knijff, Wiebe, Griffiths, Templeton, & Hammer, 1999; Underhill, Shen, Lin, Jin, Passarino, Yang, Kauffman, Bonne-Tamir, Bertranpetit, Francalacci, Ibrahim, Jenkins, Kidd, Mehdi, Seielstad, Wells, Piazza, Davis, Feldman, Cavalli-Sforza, & Oefner, 2000). The antiquity of the K-M9 lineage is consistent with the presence of this SNP in a sizeable majority of Siberian Y chromosomes (Karafet, Zegura, Posukh, Osipova, Bergen, Long, Goldman, Klitz, Harihara, de Knijff, Wiebe, Griffiths, Templeton, & Hammer, 1999; Lell, Sukernik, Starikovskaya, Su, Jin, Schurr, Underhill, & Wallace, 2002). The oldest SNP in the Eurasian branch of the Y chromosome phylogeny, F-89, dates to ~62,000 ybp (Schurr & Sherry, 2004), and predates the occurrence of the K-M9 lineage, since it appears in all haplotypes bearing the latter mutation. F-89 is an important SNP because it marks the initial diversification and spread of non-African Y chromosome lineages into the rest of the world.

By contrast, the C-M130 lineage is somewhat younger than the F-M89 or K-M9 lineages, having been dated at ~30,000 – 25,000 ybp (Karafet, Zegura, Posukh, Osipova, Bergen, Long, Goldman, Klitz, Harihara, de Knijff, Wiebe, Griffiths, Templeton, & Hammer, 1999; Underhill, Shen, Lin, Jin, Passarino, Yang, Kauffman, Bonne-Tamir, Bertranpetit, Francalacci, Ibrahim, Jenkins, Kidd, Mehdi, Seielstad, Wells, Piazza, Davis, Feldman, Cavalli-Sforza, & Oefner, 2000). Its age is generally consistent with its broad distribution in East and Southeast Asia, in which it appears to have originated, and its haplotypic diversity in eastern Siberian and Asian populations (Lell, Sukernik, Starikovskaya, Su, Jin, Schurr, Underhill, & Wallace, 2002). The estimated age of haplogroup R1a1-M17 is rather intriguing. Using the SNP evolution rate of Underhill et al. (2000), an age of 13,800 ybp was obtained for this lineage, one that falls into the very end of the Last Glacial Maximum (Schurr & Sherry, 2004). These haplotypes constitute a distinct branch within R1a, and are not especially common in Siberian populations, although occurring across a broad geographic area (Lell et al., 2002). The data suggest that R1a1-M17 haplotypes did not emerge in Siberia until after the Americas had already been colonized and were brought to the Americas through a secondary expansion of ancient Asian populations, along with C-M130 and P-M45b haplotypes (Lell et al., 2002). The most recent efforts to date the Y chromosome haplotypes present in American Indian populations have utilized a newly identified SNP, Q-M242, to make this estimate (Bortolini, Salzano, Thomas, Stuart, Nasanen, Bau, Hutz, Layrisse, Petzl-Erler, Tsuneto, Hill, Hurtado, Castro-de-Guerra, Torres, Groot, Michalski, Nymadawa, Bedoya, Bradman, Labuda, & Ruiz-Linares, 2003; Seielstad, Bekele, Ibrahim, Toure, & Traore, 1999). The Q-M242 marker occurred within haplogroup P-M45 in Central Asia prior to the emergence of the Q-M3 SNP and the expansion of its haplotypes in the Americas. As such, the M242 marker may better define the initial entry into the Americas than M3. Using STR mutation rates, researchers have dated the age of M242 haplotypes in the Americas at ~18,000 – 15,000 ybp (Bortolini et al., 2003; Seielstad et al., 1999).

The analysis of mtDNA unambiguously shows that the ancestral populations of contemporary American Indian populations originated from northeastern Asia (Horai et al., 1993; Schurr, Ballinger, Gan, Hodge, Weiss, & Wallace, 1990; Shields, Schmiechen, Frazier, Redd, Voevoda, Reed, & Ward, 1993; Torroni, Schurr, Yang, Szathmary, Williams, Schanfield, Troup, Knowler, Lawrence, & Weiss et, 1992; Ward, Frazier, Dew-Jager, & Paabo, 1991). It is now clear that multiple mitochondrial lineages have contributed to American Indian populations, whether revealed by restriction site analysis (Schurr, Ballinger, Gan, Hodge, Weiss, & Wallace, 1990; Torroni, Chen, Semino, Santachiara-Beneceretti, Scott, Lott, Winter, & Wallace, 1994; Torroni, Schurr, Yang, Szathmary, Williams, Schanfield, Troup, Knowler, Lawrence, & Weiss et, 1992), or by high resolution sequence analysis (Horai, Kondo, Nakagawa-Hattori, Hayashi, Sonoda, & Tajima, 1993; Ward, Alan Redd, Valencia, Frazier, & Paabo, 1993; Ward, Frazier, Dew-Jager, & Paabo, 1991). However, mtDNA represents only a small fraction of the human genome and a phylogeny based only on mtDNA is an imperfect estimate of the population phylogeny (Marjoram, 1994; Pamilo & Nei, 1998). Earlier studies used nuclear polymorphisms to good effect to investigate the processes underlying American Indian evolution: "classical" genetic markers emphasized the importance of intra-tribal population structure (Chakraborty, Smouse, & Neel, 1988; Neel & Thompson, 1978; Ward, 1972), with intra-tribal micro-differentiation accounting for a substantial portion of the continent wide variability (Neel & Ward, 1970). Analysis of nuclear polymorphisms demonstrated broad continental trends of genetic diversity (Cavalli-Sforza, Menozzi, & Piazza, 1993; Spuhler, 1979; Suarez, Crouse, & O'Rourke, 1985), with apparent stasis in Central America (Barrantes et al., 1990). This latter interpretation is now supported by recent mtDNA data (Torroni, 1994). Overall, despite the occasional furor about the finer points of linguistic classification (Dillehay & Collins, 1988; Greenberg, 1987; Nichols, 1990; Ruhlen, 1992) and time of genetic divergence (Torroni, 1994; Ward, Frazier, Dew-Jager, & Paabo, 1991), there are a number of consistent relationships between language, culture, and biological affinities throughout most of the Americas (Driver & Coffin, 1975; Greenberg, Turner, & Zegura, 1986; Suarez, Crouse, & O'Rourke, 1985). Hence, an intensive investigation of molecular diversity at the sequence level in American Indian populations is extremely likely to reveal information about the tempo and mode of molecular evolution in this important chapter of human history.

Based on the existing molecular anthropological data, the following picture of the peopling of the Americas is offered in conclusion. There was a pre-Clovis entry of ancestral Asian groups into the Americas during the Last Glacial Maximum. These immigrants used a coastal route to reach the areas below the glaciated areas of northern North America somewhere between 18,000–15,000 ybp. They apparently brought mtDNA haplogroups A–D and Y chromosome haplogroups P-M45a and Q-242/Q-M3 haplotypes with them to the Americas, with these being dispersed throughout all continental areas of the Americas. A subsequent expansion probably brought mtDNA haplogroup X and Y chromosome haplogroups P-M45b, C-M130, and R1a1-M17, with these being disseminated in only North and Central America. This expansion may have coincided with the opening of the ice-free corridor around 12,550 ybp. A somewhat later expansion likely involved the emergence of circumarctic populations, such as Eskimos, Aleuts, and Na-Dene Indians (Rubicz, Melvin, & Crawford, 2002; Rubicz, Schurr, Babb, & Crawford, 2003; Saillard, Forster, Lynnerup, Bandelt, & Norby, 2000; Shields, Schmiechen, Frazier, Redd, Voevoda, Reed, & Ward, 1993).

REFERENCES

Advise, C., Arnold, J., & Ball, R. M. (1987). Intraspecific Phylogeography: The Mitochondrial DNA Bridge Between Population Genetics and Systematics. Annual Review of Ecology and Systematics, 18, 489-522.

Akazawa, T. (1999). Pleistocene Peoples of Japan and the Peopling of the Americas. In Bonnichsen, R. & Turnmire, K. (Eds.), Ice Age People of North America: Environments, Origins, Adaptations (pp. 95-103). Corvallis, OR: Oregon State University Press.

Alves-Silva, J., Santos, M. d. S., Guimaraes, P. E. M., Fereira, A. C. S., Bandelt, H.-J., Pena, S. D. J., et al. (2000). The Ancestry of Brazilian mtDNA Lineages. American Journal of Human Genetics, 67, 444-461.

Anderson, D. G., & Gillam, J. C. (2001). Paleoindian Interaction and Mating Networks: Reply to Moore and Moseley. American Antiquity, 66(3), 530-535.

Anderson, S., Bankier, A. T., Barrell, B. G., De Bruijn, M. H. L., Coulson, A. R., Drouin, J., et al. (1981). Sequence and Organization of the Human Mitochondrial Genome. Nature, 290, 457-474.

Aquadro, C. F., & Greenberg, B. D. (1983). Human Mitochondrial DNA Variation and Evolution: Analysis of Nucleotide Sequences from Seven Individuals. Genetics, 103, 287-312.

Bahlo, M., & Griffiths, R. C. (2000). Inference from Gene Trees in a Subdivided Population. Theoretical Population Biology, 57, 79-95.

Bailliet, G., Rothhammer, F., Carnese, F. R., Bravi, C. M., & Bianchi, N. O. (1994). Founder mitochondrial haplotypes in Amerindian populations. Am J Hum Genet, 55(1), 27-33.

Barrantes, R., Smouse, P. E., Mohrenweiser, H. W., Gershowitz, H., Azofeifa, J., Arias, T. D., et al. (1990). Microevolution in Lower Central America: Genetic Characterization of the Chibcha Speaking Groups of Costa Rica and Panama, and a Consensus Taxonomy Based on Genetic and Linguistic Affinity. American Journal of Human Genetics, 46, 63-84.

Batista, O., Kolman, C. J., & Bermingham, E. (1995). Mitochondrial DNA diversity in the Kuna Amerinds of Panama. Hum Mol Genet, 4(5), 921-9.

Batzer, M. A., Stoneking, M., & Alegria-Hartman, M. (1994). African Origin of Human-specific Polymorphic Alu Insertions. Proceedings of the National Acadamie of Sciences, 88, 839-843.

Beaumont, M. (1999). Detecting Population Expansion and Decline Using Microsatellites. Genetics, 153, 2013-2029.

Bergen, A. W., Wang, C. Y., Tsai, J., Jefferson, K., Dey, C., Smith, K. D., et al. (1999). An Asian-Native American paternal lineage identified by RPS4Y resequencing and by microsatellite haplotyping. Annals of Human Genetics, 63, 63-80.

Bever, M. R. (2001). An Overview of Alaskan Late Pleistocene Archaeology: Historical Themes and Current Perspectives. Journal of World Prehistory, 15(2), 125-191.

Bianchi, N., Bailliet, G., Bravi, C., Carnese, R., Rothhammer, F., Martinez-Marignac, V., et al. (1997). Origin of Amerindian Y-Chromosomes as Inferred by the Analysis of Six Polymorphic Markers. American Journal of Physical Anthropology, 102, 79-89.

Bianchi, N. O., Catanesi, C. I., Bailliet, G., Martinez-Marignac, V. L., Bravi, C. M., Vidal-Rioja, L. B., et al. (1998). Characterization of Ancestral and Derived Y-Chromosome Haplotypes of New World Native Populations. American Journal of Human Genetics, 63, 1862-1871.

Bianchi, N. O., & Rothhammer, F. (1995). Reply to Torroni and Wallace. American Journal of Human Genetics, 56, 1236-1238.

Bodmer, J., & Bodmer, W. F. (1973). Population Genetics of the HLA System. A Summary of Data from the Fifth International Histocompatibility Testing Workshop. Israel Journal of Medical Science, 9, 1257-1268.

Bonatto, S., & Salzano, F. (1997a). Diversity and Age of the Four Major mtDNA Haplogroups, and Their Implications for the Peopling of the New World. American Journal of Human Genetics, 61, 1413-1423.

Bonatto, S. L., & Salzano, F. M. (1997b). A single and early migration for the peopling of the Americas supported by mitochondrial DNA sequence data. Proceedings of the National Academy of Sciences of the United States of America, 94(5), 1866-1871.

Bortolini, M. C., Salzano, F. M., Bau, C. H., Layrisse, Z., Petzl-Erler, M. L., Tsuneto, L. T., et al. (2002). Y-chromosome biallelic polymorphisms and Native American population structure. Ann Hum Genet, 66(Pt 4), 255-9.

Bortolini, M. C., Salzano, F. M., Thomas, M. G., Stuart, S., Nasanen, S. P., Bau, C. H., et al. (2003). Y-chromosome evidence for differing ancient demographic histories in the Americas. American Journal of Human Genetics, 73(3), 524-39.

Boyd, W. (1952). Genetics and the Races of Man. Boston, MA: Little, Brown and Co.

Bradman, N., & Thomas, M. (1998). Why Y? The Y Chromosome in the Story of Human Evolution, Migration, and Prehistory. Science Spectra, 14, 1-8.

Brown, M. D., Hosseini, S. H., Torroni, A., Bandelt, H.-J., Allen, J. C., Schurr, T. G., et al. (1998). mtDNA Haplogroup X: An Ancient Link between Europe/Western Asia and North America. American Journal of Human Genetics, 63, 1852-1861.

Brown, W. M., Prager, E. M., Wang, A., & Wilson, A. C. (1979). Rapid Evolution of Animal Mitochondrial DNA. Proceedings of the National Academy of Sciences, 76, 1967-1971.

Cann, R. L. (1982). The Evolution of Human Mitochondrial DNA. Unpublished Doctorial Dissertation, University of California, Berkeley, CA.

Cann, R. L. (1985). Mitochondrial DNA Variation and the Spread of Modern Populations. In Kirk, R. L. (Ed.), Out of Asia (pp. 113-122). Canberra: Astralian National University.

Cann, R. L., Stoneking, M., & Wilson, A. C. (1987). Mitochondrial DNA and Human Evolution. Nature, 325, 31-36.

Cannings, C., & Cavalli-Sforza, L. L. (1973). Human Population Structure. Advances in Human Genetics, 4, 105-171.

Carlyle, S. W., Parr, R. L., Hayes, M. G., & O'Rourke, D. H. (2000). Context of maternal lineages in the greater Southwest. American Journal of Physical Anthropology, 113(1), 85-101.

Carlyle, S. W., Ryan L. Parr, M. Geoffrey Hayes, and Dennis H. O'Rourke. (2000). Context of Maternal Lineages in the Great Southwest, American Journal of Physical Anthropology (Vol. 113:85-101).

Cavalli-Sforza, L. L. (1984). Isolation by Distance. In Chakravarti, A. (Ed.), Human Population Genetics: The Pittsburgh Symposium (pp. 229-247). New York, NY: Van Nostrand Reinhold.

Cavalli-Sforza, L. L., & Bodmer, W. F. (1971). The Genetics of Human Populations. San Francisco, CA: W.H. Freeman and Company.

Cavalli-Sforza, L. L., & Edwards, A. W. F. (1965). Analysis of Human Evolution. In Geerts, J. (Ed.), Genetics Today: Proceedings of the Eleventh International Conference of Genetics, The Hague, September 1963 (Vol. 3, pp. 923-933). New York, NY: Pergamum Press.

Cavalli-Sforza, L. L., Menozzi, P., & Piazza, A. (1993). Demic Expansions and Human Evolution. Science, 259, 639-646.

Cavalli-Sforza, L. L., Menozzi, P., & Piazza, A. (1994). The History and Geography of Human Genes. New Jersey, NJ: Princeton University Press.

Cavalli-Sforza, L. L., Piazza, A., Menozzi, P., & Mountain, J. L. (1988). Reconstruction of Human Evolution; Bringing Together Genetic, Archaeological, and Linguistic Data. Proceedings of the National Academy of Sciences, 85, 6002-6006.

Chakraborty, R., Smouse, P. E., & Neel, J. V. (1988). Population Amalgamation and Genetic Variation: Observations on Artificially Agglomerated Populations of Central and South America. American Journal of Human Genetics, 43, 709-725.

Chen, Y.-S., Olckers, A., Schurr, T. G., Kogelnik, A. M., Huoponen, K., & Wallace, D. C. (2000). mtDNA Variation in the South African Kung and Khwe- and Their Genetic Relationship to Other African Populations. American Journal of Human Genetics, 66, 1362-1383.

Consortium, Y. C. (2002). A Nomenclature System for the Tree of Y Chromosomal Binary Haplogroups. Genome Research, 12, 339-348.

Crawford, M. H. (1973). The Use of Genetic Markers of the Blood in the Study of the Evolution of Human Populations. In Workman, P. L. (Ed.), Methods and Theories of Anthropological Genetics (pp. 19-38). Albuquerque, NM: University of New Mexico Press.

Crow, J. F., & Kimura, M. (1970). An Introduction to Population Genetics Theory. New York, NY: Harper & Row.

Dillehay, T. D., & Collins, M. B. (1988). Early Cultural Evidence from Monte Verde in Chile. Nature, 332, 150-152.

Donnelly, P., & Tavare, S. (1995). Coalescents and Genealogical Structure Under Neutrality. Annual Review of Genetics, 29, 401-421.

Driver, H. E., & Coffin, J. L. (1975). Classification and Development of North American Indian Cultures: A Statistical Analysis of the Driver-Massey Sample. Transactions of the American Philosophical Society, 65, 1-32.

Easton, R. D., Merriwether, A., Crews, D. E., & Ferrell, R. E. (1996). mtDNA Variation in the Yanomami: Evidence for Additional New World Founding Lineages. American Journal of Human Genetics, 59, 213-225.

Epperson, B. K. (1994). Spatial and Space-Time Correlations in Systems of Subpopulations with Stochastic Migration. Theoretical Population Biology, 46, 160-197.

Erdtmann, B., Salzano, F. M., & Mattevi, M. S. (1981). Size variability of the Y chromosome distal C-band in Brazilian Indians and Caucasoids. Ann Hum Biol, 8(5), 415-24.

Fearnhead, P., & Donnelly, P. (2001). Estimating Recombination Rates from Population Genetics Data. Genetics, 159, 1299-1318.

Felsenstein, J. (1975). A Pain in the Torus: Some Difficulties with Models of Isolation by Distance. American Naturalist, 109, 359-368.

Fix, A. G. (1979). Anthropological Genetics of Small Populations. Annual Review of Anthropology, 8, 207-230.

Forster, P., Harding, R., Torroni, A., & Bandelt, H.-J. (1996). Origin and Evolution of Native American mtDNA Variation: A Reappraisal. American Journal of Human Genetics, 59, 935-945.

Fox, C. L. (1996). Mitochondrial DNA haplogroups in four tribes from Tierra del Fuego Patagonia: Inferences about the peopling of the Americas. Human Biology, 68(6), 855-871.

Freimer, N., & Slatkin, M. (1996). Microsatellites: Evolution and Mutational Processes. Paper presented at the Symposium on Variation in the Human Genome, Ciba Foundation, London.

Greenberg, J. (1987). Language in the Americas. San Jose, CA: Stanford University Press.

Greenberg, J., Turner, C. G. I., & Zegura, S. (1986). The settlement of the Americas: A comparison of the linguistic, dental, and genetic evidence. Current Anthropology, 27, 477-497.

Griffiths, R. C., & Marjoram, P. (1996). Ancestral Inference from Samples of DNA Sequences with Recombination. Journal of Computational Biology, 3, 479-502.

Griffiths, R. C., & Tavare, S. (1994a). Ancestral Inference in Population Genetics. Statistical Science, 9, 307-319.

Griffiths, R. C., & Tavare, S. (1994b). Sampling Theory for Neutral Alleles in a Varying Environment. Philosophical Transactions of the Royal Society of London Series B-Biological Sciences, 344, 403-410.

Griffiths, R. C., & Tavare, S. (1994c). Simulating Probability Distributions in the Coalescent. Theoretical Population Biology, 46, 131-159.

Haines, F. (1938a). The Northward Spread of Horses Among the Plains Indians. American Anthropologist, 40(3), 429-437.

Haines, F. (1938b). Where Did the Plains Indians get their Horses? American Anthropologist, 40(1), 112-117.

Hammer, M. F., A.B. Spurdle, T. Karafet, M.R. Bonner, E.T. Wood, A. Novelletto, P. Malaspina, R.J. Mitchell, S. Horai, T. Jenkins, and S.L. Zegura. (1997). The Geographic Distribution of Human Y Chromosome Variation, Genetics (Vol. 145:787-805).

Hammer, M. F., T. Karafet, A. Rasanayagam, E.T. Wood, T.K. Altheide, T. Jenkins, R.C. Griffiths, A.R. Templeton, and S.L. Zegura. (1998). Out of Africa and Back Again: Nested Cladistic Analysis of Human Y Chromosome Variation, Molecular Biological Evolution (Vol. 15(4):427-441).

Hayden, B. (Ed.). (1992). A Complex Culture of the British Columbia Plateau: Traditional Stl'atl'imx Resource Use. Vancouver, B.C.: UBC Press.

Higuchi, R. H., Bowman, B., Freiberger, M., Ryder, O. A., & Wilson, A. C. (1984). DNA Sequences from the Quagga, an Extinct Member of the Horse Family. Nature, 312, 282-284.

Hoelzer, G. A., Wallman, J., & Melnick, D. J. (1998). The Effects of Social Structure, Geographical Structure, and Population Size on the Evolution of Mitochondrial DNA: II. Molecular Clocks and the Lineage Sorting Period. Journal of Molecular Evolution, 47, 21-31.

Horai, S., Kondo, R., Nakagawa-Hattori, Y., Hayashi, S., Sonoda, S., & Tajima, K. (1993). Peopling of the Americas, founded by four major lineages of mitochondrial DNA. Molecular Biological Evolution, 10(1), 23-47.

Hudson, R. R. (1990). Gene Genealogies and the Coalescent Process. In Futuyma, D. & Antonovics, J. (Eds.), Oxford Surveys in Evolutionary Biology (Vol. 7, pp. 1-44). Oxford, UK: University of Oxford Press.

Ikawa-Smith, F. (1982). The Early Prehistory of the Americas as Seen From Northeast Asia. In Ericson, J. E., Taylor, R. E. & Berger, R. (Eds.), Peopling of the New World (pp. 13-34). Menlo Park, CA: Ballena Press.

Jacquard, A. (1975). Inbreeding: One Word, Several Meanings. Theoretical Population Biology, 7, 338-363.

Jeffreys, A. J., Wilson, V., & Their, S. L. (1985). Individual-specific "fingerprints" of human DNA. Nature, 316, 75-79.

Johnson, M. J., Wallace, D., Ferris, S. D., Rattazzi, M. C., & Cavalli-Sforza, L. L. (1983). Radiation of Human Mitochondria DNA Types Analyzed by Restriction Endonuclease Cleavage Patterns. Journal of Molecular Evolution, 19, 255-271.

Jones, P. (2002). American Indian Demographic History and Cultural Affiliation: A Discussion of Certain Limitations on the Use of mtDNA and Y Chromosome Testing. AnthroGlobe Journal, Winter, 1-32.

Jones, P. (2004). American Indian Cultural Affiliation and Cultural Continuity in the Plateau and Great Basin Culture Regions of the American West. Saybrook Graduate School, San Francisco.

Jorde, L. B. (1980). The Genetic Structure of Subdivided Human Populations: A Review. In Mielke, J. H. & Crawford, M. H. (Eds.), Current Developments in Anthropological Genetics (pp. 135-208). New York, NY: Plenum.

Jorde, L. B. (1985). Human Genetic Distance Studies: Present Status and Future Prospects. Annual Review of Anthropology, 14(1), 343-373.

Jorde, L. B., Bamshad, M., & Rogers, A. R. (1998). Using Mitochondrial and Nuclear DNA Markers to Reconstruct Human Evolution. BioEssays, 20, 126-136.

Kaestle, F. (1995). Mitochondrial DNA Evidence for the Identity of the Descendants of the Prehistoric Stillwater Marsh Populations. In Larsen, C. S. & Kelly, R. L. (Eds.), Bioarchaeology of the Stillwater Marsh: Prehistoric Human Adaptation in the Western Great Basin (Vol. 77, pp. 73-80). Washington, D.C.: American Museum of Natural Hisotry.

Kaestle, F. (1997). Molecular Analysis of Ancient Native American DNA from Western Nevada. Nevada Historical Society Quarterly, 40(1), 85-96.

Kaestle, F. (1998). Molecular Evidence for Prehistoric Native American Population Movement: The Numic Expansion. Unpublished Dissertation, University of California, Davis, Davis, CA.

Kaestle, F. (2000a). Comment on the Repatriation of the "Spirit Cave Man" Remains. Reno, NV: Ms on file in BLM Nevada office.

Kaestle, F. (2000b). Report on DNA Analysis of the Remains of "Kennewick Man" from Columbia Park, Washington: Department of the Interior: Washington D.C.

Kaestle, F. A., & Smith, D. G. (2001a). Ancient mitochondrial DNA evidence for prehistoric population movement: The Numic expansion. American Journal of Physical Anthropology, 115(1), 1-12.

Kaestle, F. A., & Smith, D. G. (2001b). Ancient Mitochonrial DNA Evidence for Prehistoric Population Movement: The Numic Expansion. American Journal of Physical Anthropology, 115, 1-12.

Karafet, T., de Knijff, P., & Wood, E. (1998). Different patterns of variation at the X- and Y-chromosome-linked microsatellite loci DXYS156X and DXYS156Y in human populations. Human Biology.

Karafet, T., Zegura, S. L., Vuturo-Brady, J., Posukh, O., Osipova, L., Wiebe, V., et al. (1997). Y Chromosome Markers and Trans-Bering Strait Dispersals. American Journal of Physical Anthropology, 102, 301-314.

Karafet, T. M., Zegura, S. L., Posukh, O., Osipova, L., Bergen, A., Long, J., et al. (1999). Ancestral Asian Source(s) of New World Y-Chromosome Founder Haplotypes. American Journal of Human Genetics, 64, 817-831.

Kayser, M., Brauer, S., Weiss, G., Schiefenhovel, W., Underhill, P. A., & Stoneking, M. (2001). Independent Histories of Human Y Chromosomes from Melanesia and Australia. American Journal of Human Genetics, 68, 173-190.

Kayser, M., Krawczak, M., Excoffier, L., Dieltjes, P., Corach, D., Pascali, V., et al. (2001). An Extensive Analysis of Y-Chromosomal Microsatellite Haplotypes in Globally Dispersed Human Populations. American Journal of Human Genetics, 68, 990-1018.

Kimura, M. (1968). Evolutionary Rate at the Molecular Level. Nature, 217, 624-626.

Kimura, M., & Weiss, G. H. (1964). The Stepping Stone Model of Population Structure and the Decrease of Genetic Correlation with Distance. Genetics, 49, 561-576.

Kingman, J. F. C. (1982). On the Genealogy of Large Populations. In Hannan, E. J. (Ed.), Essays in Statistical Science: Papers in Honor of P.A.P. Moran (pp. 27-43). Sheffield, UK: Applied Probability Trust.

Kuhner, M. K., Yamato, J., & Felsenstein, J. (1995). Estimating Effective Population Size and Mutation Rate from Sequence Data using Metropolis-Hastings Sampling. Genetics, 140, 1421-1430.

Lalueza, C., PerezPerez, A., Prats, E., Cornudella, L., & Turbon, D. (1997). Lack of founding Amerindian mitochondrial DNA lineages in extinct aborigines from Tierra del Fuego Patagonia. Human Molecular Genetics, 6(1), 41-46.

Lalueza-Fox, C., Calderon, F. L., Calafell, F., Morera, B., & Bertranpetit, J. (2001). MtDNA from extinct Tainos and the peopling of the Caribbean. Annals of Human Genetics, 65, 137-151.

Lalueza-Fox, C., Gilbert, M. T., Martinez-Fuentes, A. J., Calafell, F., & Bertranpetit, J. (2003). Mitochondrial DNA from pre-Columbian Ciboneys from Cuba and the prehistoric colonization of the Caribbean. Am J Phys Anthropol, 121(2), 97-108.

Lampl, M., & Blumberg, B. S. (1979). Blood Polymorphisms and the Origin of New World Populations. In Harper, A. B. (Ed.), The First Americans: Origins, Affinities and Adaptations. New York, NY: Gustav Fischer.

Layrisse, M. (1968). Biological Subdivisions of the Indian on the Basis of Genetic Traits. In Biomedical Challenges Presented by the American Indian (pp. 35-39). Washington, DC: Wold American Health Organization.

Lell, J. T. (2000). Y chromosome analysis of Native American and Siberian populations : evidence for two independent migrations of New World male founders.

Lell, J. T., Brown, M. D., Schurr, T. G., Sukernik, R. I., Starikovskaya, Y. B., Torroni, A., et al. (1997). Y chromosome polymorphisms in Native American and Siberian populations: Identification of Native American Y chromosome haplotypes. Human genetics, 100(5-6), 8.

Lell, J. T., Sukernik, R. I., Starikovskaya, Y. B., Su, B., Jin, L., Schurr, T. G., et al. (2002). The dual origin and Siberian affinities of Native American Y chromosomes. American Journal of Human Genetics, 70(1), 192-206.

Levins, R. (1966). The Strategy of Model Building in Population Biology. American Scientist, 54, 421-431.

Lin, S. J., Tanaka, K., Leonard, W. R., Gerelsaikhan, T., Dashnyam, B., Nymakhishig, S., et al. (1994). A Y-associated Allele is Shared Among a Few Ethnic Groups of Asia. Japanese Journal of Human Genetics, 39, 299-304.

Livingstone, F. B. (1989). Simulation of the Diffusion of the B-globin variants in the Old World. Human Biology, 61, 267-309.

Lorenz, J., & Smith, D. G. (1996). Distribution of Four Founding mtDNA Haplogroups Among Native North Americans. American Journal of Physical Anthropology, 101, 307-323.

Lorenz, J. G., & Smith, D. G. (1997). Distribution of sequence variation in the mtDNA control region of native North Americans. Human Biology, 69(6), 749-776.

MacCluer, J. W., & Dyke, B. (1976). On the Minimum Size of Endogamous Populations. Social Biology, 23, 1-12.

MacEachern, S. (2000). Genes, Tribes, and African History. Current Anthropology, 41(3), 357-384.

Malecot, G. (1955). The Decrease of Relationship with Distance. Cold Springs Harbor Symposium on Quantitative Biology, 20, 52-53.

Malecot, G. (1967). Identical Loci and Relationship. Proceeding of the Fifth Berkeley Symposium of Mathimatical Statistical Probabilities, 4, 317-332.

Malecot, G. (1973). Isolation by Distance. In Morton, N. E. (Ed.), Genetic Structure of Populations. Honolulu, HI: University Press of Hawaii.

Malecot, G. (1975). Heterozygosity and Relationship in Regularly Subdivided Populations. Theoretical Population Biology, 8, 212-241.

Malhi, R. S., Eshleman, J. A., Greenberg, J. A., Weiss, D. A., Schultz Shook, B. A., Kaestle, F. A., et al. (2002). The structure of diversity within New World mitochondrial DNA haplogroups: implications for the prehistory of North America. Am J Hum Genet, 70(4), 905-19.

Malhi, R. S., Schultz, B. A., & Smith, D. G. (2001). Distribution of mitochondrial DNA lineages among native American tribes of northeastern North America. Human Biology, 73(1), 17-55.

Malhi, R. S., & Smith, D. G. (2002). Brief communication: Haplogroup X confirmed in prehistoric North America. American Journal of Physical Anthropology, 119(1), 84-86.

Marjoram, P., and P. Donnelly. (1994). Pairwise Comparisons of Mitochondrial DNA Sequences in Subdivided Populations and Implications for Early Human Evolution, Genetics (Vol. 136:673-683).

Markovtsova, L., Marjoram, P., & Tavare, S. (2000a). The age of a unique event polymorphism. Genetics, 156(1), 401-9.

Markovtsova, L., Marjoram, P., & Tavare, S. (2000b). The Effects of Rate Variation on Ancestral Inference in the Coalescent. Genetics, 156, 1427-1436.

Matsumoto, H., & Miyazaki, T. (1972). Gm and Inv Allotypes of the Ainu in Hidaka area Hokkaido. Japanese Journal of Human Genetics, 17, 20-26.

Merriwether, A., & Cabana, G. (2000). Kennewick Man Ancient DNA Analysis: Draft Report Submitted to the Department of the Interior, National Park Service. Washington, D.C.: Department of the Interior.

Merriwether, A., & Ferrell, R. E. (1996). The Four Founding Lineage Hypothesis for the New World: A Critical Reevaluation. Molecular Phylogenetics and Evolution, 5(1), 241-246.

Merriwether, A., Hell, W. W., Vahlne, A., & Ferrell, R. E. (1996). mtDNA Variation Indicates Mongolia May Have Been the Source for the Founding Population for the New World. American Journal of Human Genetics, 59, 204-212.

Merriwether, A., Rothhammer, F., & Ferrell, R. E. (1995). Distribution of the Four Founding Lineage Haplotypes in Native Americans Suggests a Single Wave of Migration for the New World. American Journal of Physical Anthropology, 98, 411-430.

Merriwether, D. A. (2000). Report on DNA Analysis of the Kennewick Human Remains (Draft). Washington, D.C.: Department of the Interior.

Merriwether, D. A., Huston, S., Iyengar, S., Hamman, R., Norris, J. M., Shetterly, S. M., et al. (1997). Mitochondrial Versus Nuclear Admixture Estimates Demonstrate a Past History of Directional Mating. American Journal of Physical Anthropology, 102, 153-159.

Merriwether, D. A., Rothhammer, F., & Ferrell, R. E. (1994). Genetic variation in the New World: ancient teeth, bone, and tissue as sources of DNA. Experientia, 50(6), 592-601.

Monsalve, M. V., Edin, G., & Devine, D. V. (1998). Analysis of HLA class I and class II in Na-Dene and Amerindian populations from British Columbia, Canada. Human Immunology, 59(1), 48-55.

Monsalve, M. V., Groot de Restrepo, H., Espinel, A., Correal, G., & Devine, D. V. (1994). Evidence of mitochondrial DNA diversity in South American aboriginals. Ann Hum Genet, 58 (Pt 3), 265-73.

Monsalve, M. V., & Hagelberg, E. (1997). Mitochondrial DNA polymorphisms in Carib people of Belize. Proc R Soc Lond B Biol Sci, 264(1385), 1217-24.

Morton, N. E. (1972). The Future of Human Population Genetics. Progress in Medical Genetics, 8, 103-124.

Morton, N. E. (1977). Isolation by Distance in Human Populations. Annals Of Human Genetics, 40, 361-365.

Morton, N. E. (Ed.). (1973). Genetic Structure of Populations. Honolulu: HI: University of Hawaii Press.

Mountain, J. L., & Cavalli-Sforza, L. L. (1997). Multilocus Genotypes, a Tree of Individuals, and Human Evolutionary History. American Journal of Human Genetics, 61, 705-718.

Mullis, K. B., & Faloona, F. (1987). Specific Synthesis of DNA in Vitro via a Polymerase-Catalysed Chain Reaction. Methods Enzymology, 155, 335-350.

Neel, J. V., & Salzano, F. M. (1967). Further Studies on the Xavante Indians. X. Some Hypotheses-Generalizations Resulting from these Studies. American Journal of Human Genetics, 19, 554-574.

Neel, J. V., & Thompson, E. A. (1978). Founder Effect and the Number of Private-Polymorphisms Observed in Amerindian Tribes. Proceedings of the National Acadamie of Sciences, 75, 1904-1908.

Neel, J. V., & Ward, R. H. (1970). Village and Tribal Genetic Distances Among American Indians and the Possible Implications for Human Evolution. Proceedings of the National Acadamie of Sciences, 65, 323i-330.

Nei, M. (1987). Molecular Evolutionary Genetics. Columbia, PA: Columbia University Press.

Nichols, J. (1990). Linguistic Diversity and the First Settlement of the New World. Language, 66, 475-521.

Nielsen, R. (1997). A Likelihood Approach to Population Samples of Microsatellite Alleles. Genetics, 146, 711-716.

O'Rourke, D. H., Hayes, M. G., & Carlyle, S. W. (2000a). Ancient DNA studies in physical anthropology. Annual Review of Anthropology.

O'Rourke, D. H., Hayes, M. G., & Carlyle, S. W. (2000b). Spatial and temporal stability of mtDNA haplogroup frequencies in native North America. Human Biology.

O'Rourke, D. H., Mobarry, A., & Suarez, B. K. (1992). Patterns of Genetic Variation in Native America. Human Biology, 64(3), 417-434.

Paabo, S. (1985a). Molecular Cloning of Ancient Egyptian Mummy DNA. Nature, 314, 644-645.

Paabo, S. (1985b). Preservation of DNA in Ancient Egyptian Mummies. Journal of Archaeological Science, 12, 411-417.

Paabo, S., Gifford, J. A., & Wilson, A. C. (1988). Mitochondrial DNA Sequences from a 7000-year-old brain. Nucleic Acid Research, 16, 9775-9787.

Page, R. D. M., & Charleston, M. A. (1990). Reconciled Trees and Incongruent Gene and Species Trees. DIMACS Series in Discrete Mathematics and Theoretical Computer Science, 00, 1-14.

Pamilo, P., & Nei, M. (1998). Relationships Between Gene Trees and Species Trees. Molecular Biological Evolution, 5(5), 568-583.

Parr, R. L., Carlyle, S. W., & ORourke, D. H. (1996). Ancient DNA analysis of Fremont Amerindians of the Great Salt Lake Wetlands. American Journal of Physical Anthropology, 99(4), 507-518.

Pena, S. D., Santos, F. R., Bianchi, N. O., Bravi, C. M., Carnese, F. R., Rothhammer, F., et al. (1995). A major founder Y-chromosome haplotype in Amerindians. Nat Genet, 11(1), 15-6.

Poinar, H. N., Kuch, M., Sobolik, K. D., Barnes, I., Stankiewicz, A. B., Kuder, T., et al. (2001). A molecular analysis of dietary diversity for three archaic Native Americans. Proceedings Of The National Academy Of Sciences Of The United States Of America, 98(8), 4317-4322.

Renfrew, C., & Boyle, K. (Eds.). (2000). Archaeogenetics: DNA and the Population Prehistory of Europe. Cambridge, UK: McDonald Institute for Archaeological Research.

Ribeiro-dos-Santos, A. K., Guerreiro, J. F., Santos, S. E., & Zago, M. A. (2001). The split of the Arara population: comparison of genetic drift and founder effect. Hum Hered, 51(1-2), 79-84.

Ribeiro-dos-Santos, A. K., Santos, S. E., Machado, A. L., Guapindaia, V., & Zago, M. A. (1997). Reply to Monsalve on "Mitochondrial DNA in ancient Ameridians," American Journal of Physical Anthropology (1997) 103:423-425. Am J Phys Anthropol, 103(4), 571.

Ribiero-Dos-Santos, A. K. C., Santos, S. E. B., Machado, A. L., Guapindaia, V., & Zago, M. A. (1996). Heterogeneity of Mitochondrial DNA Haplotypes in pre-Columbian Natives of the Amazon Region. American Journal of Physical Anthropology, 101, 29-37.

Rickards, O., Martinez-Labarga, C., Lum, J. K., De Stefano, G. F., & Cann, R. L. (1999). mtDNA history of the Cayapa Amerinds of Ecuador: Detection of additional founding lineages for the native American populations. American Journal of Human Genetics, 65(2), 519-530.

Rogers, A., & Jorde, L. B. (1995). Genetic Evidence on Modern Human Origins. Human Biology, 67, 1-36.

Rogers, A. R. (1988). Three Components of Genetic Drift in Subdivided Populations. American Journal of Physical Anthropology, 77, 435-450.

Rogers, L. A., Rogers, R. A., & Martin, L. D. (1992). How the Door Opened: The Peopling of the New World. Human Biology, 64(3), 281-302.

Rubicz, R., Melvin, K. L., & Crawford, M. H. (2002). Genetic evidence for the phylogenetic relationship between Na-Dene and Yeniseian speakers. Human Biology, 74(6), 743-760.

Rubicz, R., Schurr, T. G., Babb, P. L., & Crawford, M. H. (2003). Mitochondrial DNA variation and the origins of the Aleuts. Human Biology, 75(6), 809-35.

Ruhlen, M. (1992). The Amerind Phylum and Prehistory of the New World. In Lamp, S. L. & Mitchell, E. D. (Eds.), Sprung from Some Common Source: The Prehistory of Languages (pp. 330-350). Stanford, CA: Stanford University Press.

Ruiz-Linares, A., Ortiz-Barrientos, D., Figueroa, M., Mesa, N., Munera, J. G., Bedoya, G., et al. (1999). Microsatellites provide evidence for Y chromosome diversity among the founders of the New World. Proceedings of the National Academy of Sciences, 96(11), 6312-7.

Saiki, R. K., Gelfand, D. H., Stoffel, S., Scharf, S. J., & Higuchi, R. H. (1988). Primer-directed Enzymatic Amplification of DNA with a Thermostable DNA Polymerase. Science, 239, 487-491.

Saillard, J., Forster, P., Lynnerup, N., Bandelt, H.-J., & Norby, S. (2000). mtDNA Variation among Greenland Eskimos: The Edge of the Beringian Expansion. American Journal of Human Genetics, 67, 718-726.

Salzano, F. M. (1982). The microevolutionary process—a view from South America. Acta Anthropogenet, 6(1), 1-21.

Santos, F. R., Hutz, M. H., Coimbra, C. E. A., Santos, R. V., Salzano, F. M., & Peena, S. D. J. (1995). Further Evidence for the Existence of a Major Founder Y Chromosome Haplotype in Amerindians. Brazilian Journal of Genetics, 18, 669-672.

Santos, F. R., Pandya, A., Tyler-Smith, C., Pena, S. D., Schanfield, M., Leonard, W. R., et al. (1999a). The central Siberian origin for native American Y chromosomes. American Journal Of Human Genetics, 64(2), 619-628.

Santos, F. R., Pandya, A., Tyler-Smith, C., Pena, S. D. J., Schanfield, M., Leonard, W. R., et al. (1999b). The Central Siberian Origin for Native American Y Chromosomes. American Journal of Human Genetics, 64, 619-628.

Santos, F. R., Rodriguez-Delfin, L., Pena, S. D., Moore, J., & Weiss, K. M. (1996). North and South Amerindians may have the same major founder Y chromosome haplotype. Am J Hum Genet, 58(6), 1369-70.

Schanfield, M., Crawford, M. H., Dossetor, J. B., & Gershowitz, H. (1990). Immunoglobulin Allotypes in Several North American Eskimo Populations. Human Biology, 62, 773-789.

Schurr, T., Ballinger, S., Gan, Y.-Y., Hodge, J. A., Weiss, K. M., & Wallace, D. (1990). Amerindian Mitochondrial DNAs Have Rare Asian Mutations at High Frequencies, Suggesting They Derived from Four Primary Maternal Lineages. American Journal of Human Genetics, 46, 613-623.

Schurr, T., & Wallace, D. (1999). MtDNA variation in Native Americans and Siberians and its implications for the peopling of the New World. In Bonnichsen, R. (Ed.), Who Were the First Americans? Proceedings of the 58th Annual Biology Colloquium, Oregon State University (pp. 41-77). Covallis, OR: Center for the Study of the First Americans.

Schurr, T. G., & Sherry, S. T. (2004). Mitochondrial DNA and Y Chromosome Diversity and the Peopling of the Americas: Evolutionary and Demographic Evidence. American Journal of Human Biology, 16, 420-439.

Schurr, T. G., & Wallace, D. C. (2002). Mitochondrial DNA Diversity in Southeast Asian Populations. Human Biology.

Schuster, H. (1998). Yakima and Neighboring Groups. In Walker, D. E., Jr. (Ed.), Plateau (Vol. 12, pp. 327-351). Washington, D.C.: Smithsonian Institution.

Scozzari, R., Cruciani, F., Santolamazza, P., Malaspina, P., Torroni, A., Sellitto, D., et al. (1999). Combined Use of Biallelic and Microsatellite Y-Chromosome Polymorphisms to Infer Affinities among African Populations. American Journal of Human Genetics, 65, 829-846.

Scozzari, R., Cruciani, F., Santolamazza, P., Sellitto, D., Cole, D. E., Rubin, L. A., et al. (1997). mtDNA and Y chromosome-specific polymorphisms in modern Ojibwa: implications about the origin of their gene pool. Am J Hum Genet, 60(1), 241-4.

Seielstad, M., Bekele, E., Ibrahim, M., Toure, A., & Traore, M. (1999). A View of Modern Human Origins from Y Chromosome Microsatellite Variation. Genome Research, 9, 558-567.

Shields, E. D. (1996). Quantitative complete tooth variation among East Asians and Native Americans: Developmental biology as a tool for the assessment of human divergence. Journal of Craniofacial Genetics and Developmental Biology, 16(4), 193-207.

Shields, G. F., Schmiechen, A. M., Frazier, B. L., Redd, A., Voevoda, M. I., Reed, J. K., et al. (1993). mtDNA sequences suggest a recent evolutionary divergence for Beringian and northern North American populations. American Journal Of Human Genetics, 53(3), 549-562.

Silva, W. A., Jr, Bonatto, S. L., Holanda, A. J., Ribeiro-Dos-Santos, A. K., Paixao, B. M., Goldman, G. H., et al. (2002). Mitochondrial genome diversity of Native Americans supports a single early entry of founder populations into America. American Journal Of Human Genetics, 71(1), 187-192.

Slade, P. (2000a). Most Recent Common Ancestor Probability Distribution in Gene Genealogies Under Selection. Theoretical Population Biology, 58, 291-305.

Slade, P. (2000b). Simulation of Selected Genealogies. Theoretical Population Biology, 57, 35-49.

Slatkin, M., & Hudson, R. R. (1991). Pairwise Comparisons of Mitochondrial DNA Sequences in Stable and Exponentially Growing Populations. Genetics, 129, 555-562.

Smith, D. G., Malhi, R., Eshleman, J., Lorenz, J. G., & Kaestle, F. A. (1999). Distribution of mtDNA Haplogroup X Among Native North Americans. American Journal of Physical Anthropology, 110, 271-284.

Soodyall, H., Vigilant, L., Hill, A. V., Stoneking, M., & Jenkins, T. (1996). mtDNA Control-region Sequence Variation Suggests Multiple Independent Origins of an "Asian-specific" 9-bp Deletion in Sub-Saharan Africans. American Journal of Human Genetics, 58, 595-608.

Spieth, P. T. (1974). Gene Flow and Genetic Differentiation. Genetics, 78, 961-965.

Spuhler, J. S. (1979). Genetic Distances, Trees and Maps of North American Indians. In Laughlin, W. S. & Harper, A. B. (Eds.), The First Americans: Origins, Affinities and Adaptation (pp. 135-183). New York, NY: Fischer.

Starikovskaya, Y. B., Sukernik, R. I., Schurr, T. G., Kogelnik, A. M., & Wallace, D. C. (1998). mtDNA diversity in Chukchi and Siberian Eskimos: Implications for the genetic history of ancient Beringia and the peopling of the New World. American Journal of Human Genetics, 63(5), 1473-1491.

Stern, T. (1998). Columbia River Trade Network. In Walker, D. E., Jr. (Ed.), Plateau (Vol. 12, pp. 641-652). Washington, D.C.: Smithsonian Institution.

Stone, A. C., & Stoneking, M. (1993). Ancient DNA from a pre-Columbian Amerindian Population. American Journal of Physical Anthropology, 92, 463-471.

Stone, A. C., & Stoneking, M. (1998). mtDNA Analysis of a Prehistoric Oneota Population: Implications for the Peopling of the New World. American Journal of Human Genetics, 62, 1153-1170.

Suarez, B. K., Crouse, J. D., & O'Rourke, D. H. (1985). Genetic Variation in North American Populations: The Geography of Gene Frequencies. American Journal of Physical Anthropology, 67, 217-232.

Swedlund, A. (1980). Historical Demography: Applications in Anthropological Genetics. In Mielke, J. H. & Crawford, M. H. (Eds.), Current Developments in Anthropological Genetics. New York, NY: Plenum.

Szathmary, E. J. (1993). Genetics of Aboriginal North Americans. Evolutionary Anthropology, 1, 202-220.

Takahata, N., & Satta, Y. (1997). Evolution of the Primate Lineage Leading to Modern Humans: Phylogenetic and Demographic Inferences from DNA Sequences. Proceedings of the National Academy of Sciences, 94, 4811-4815.

Takahata, N., Satta, Y., & Klein, J. (1995). Divergence Time and Population Size in the Lineage Leading to Modern Humans. Theoretical Population Biology, 48, 198-221.

Tarazona-Santos, E., & Santos, F. R. (2002). The peopling of the Americas: a second major migration? Am J Hum Genet, 70(5), 1377-80; author reply 1380-1.

Templeton, A. R. (1993). The "Eve" Hypotheses: A Genetic Critique and Reanalysis. American Anthropologist, 95(1), 51-72.

Templeton, A. R. (1998). Human races: a genetic and evolutionary perspective. American Anthropologist.

Templeton, A. R. (2002). Out of Africa Again and Again, Nature (Vol. 416:45-51).

Torroni, A., Chen, Y. S., Scott, R. C., Semino, O., Santachiarabenerecetti, S., Lott, M. T., et al. (1993). Mitochondrial-DNA and Y-Chromosome Polymorphisms in 4 Native-American Populations from Southern Mexico. American Journal of Human Genetics, 53(3), 72-72.

Torroni, A., Chen, Y. S., Semino, O., Santachiara-Beneceretti, A. S., Scott, C. R., Lott, M. T., et al. (1994). mtDNA and Y-chromosome polymorphisms in four Native American populations from southern Mexico. American Journal Of Human Genetics, 54(2), 303-318.

Torroni, A., Huoponen, K., Francalacci, P., Petrozzi, M., Morelli, L., Scozzari, R., et al. (1996). Classification of European mtDNAs from an Analysis of Three European Populations. Genetics, 144, 1835-1850.

Torroni, A., Schurr, T. G., Cabell, M. F., Brown, M. D., Neel, J. V., Larsen, M., et al. (1993). Asian affinities and continental radiation of the four founding Native American mtDNAs. American Journal Of Human Genetics, 53(3), 563-590.

Torroni, A., Schurr, T. G., Yang, C. C., Szathmary, E. J., Williams, R. C., Schanfield, M. S., et al. (1992). Native American mitochondrial DNA analysis indicates that the Amerind and the Nadene populations were founded by two independent migrations. Genetics, 130(1), 153-162.

Torroni, A., Schurr, T. G., Yang, C. C., Szathmary, E. J., Williams, R. C., Schanfield, M. S., et al. (1992). Native American mitochondrial DNA analysis indicates that the Amerind and the Nadene populations were founded by two independent migrations. Genetics, 130(1), 153-62.

Torroni, A., Sukernik, R. I., Schurr, T. G., Starikorskaya, Y. B., Cabell, M. F., Crawford, M. H., et al. (1993). mtDNA variation of aboriginal Siberians reveals distinct genetic affinities with Native Americans. American Journal Of Human Genetics, 53(3), 591-608.

Torroni, A., Theodore G. Schurr, Chi-Chuan Yang, Emoke J.E. Szathmary, Robert C. Williams, Moses S. Schanfield, Gary A. Troup, William C. Knowler, Dale N. Lawrence, Kenneth M. Weiss, and Douglas C. Wallace. (1994). Mitochondrial DNA "clock" for the Amerinds and its Implications for Timing Their Entry into North America, Proceedings of the National Acadamy of Science (Vol. 91:1158-1162).

Torroni, A., & Wallace, D. C. (1995). MtDNA haplogroups in Native Americans. American Journal Of Human Genetics, 56(5), 1234-1238.

Turnbull, C. M. (1968). The Importance of Flux in Two Hunting Societies. In Lee, R. B. & DeVore, I. (Eds.), Man the Hunter (pp. 12-32). Chicago, IL: Aldine.

Tuross, N., & Kolman, C. J. (2000). Potential for DNA Analysis of the Human Remains from Columbia Park, Kennewick, Washington. Washington, D.C.: Department of the Interior and the Department of Justice.

Underhill, P. A., Jin, L., Zemans, R., Oefner, P. J., & Cavalli-Sforza, L. L. (1996). A pre-Columbian Y chromosome-specific transition and its implications for human evolutionary history. Proceedings Of The National Academy Of Sciences Of The United States Of America, 93(1), 196-200.

Underhill, P. A., Passarino, G., Lin, A. A., Shen, P., Lahr, M. M., Foley, R. A., et al. (2001). The Phylogeography of Y Chromosome Binary Haplotypes and the Origins of Modern Human Populations. Annuals of Human Genetics, 65, 43-62.

Underhill, P. A., Shen, P., Lin, A. A., Jin, L., Passarino, G., Yang, W. H., et al. (2000). Y Chromosome Sequence Variation and the History of Human Populations. Nature Genetics, 26, 358-361.

Wainscoat, J. S., Hill, A. V., Boyce, A. L., Flint, J., Hernandez, M., Thein, S. L., et al. (1986). Evolutionary Relationships of Human Populations from an Analysis of Nuclear DNA Polymorphisms. Nature, 319, 491-493.

Wallace, D. C., Garrison, K., & Knowler, W. C. (1985). Dramatic founder effects in Amerindian mitochondrial DNAs. American Journal Of Physical Anthropology, 68(2), 149-155.

Wallace, D. C., Ye, J., Neckelmann, S. N., Singh, G., Webster, K. A., & Greenberg, B. D. (1987). Sequence Analysis of cDNAs for the Human and Bovine ATP Synthase Beta Subunit: Mitochondrial Genes Sustain Seventeen Times more Mutations. Current Genetics, 12, 81-90.

Walpoff, M. (1999). Paleoanthropology (2 ed.). New York, NY: McGraw-Hill.

Ward, R. H. (1972). The Genetic Structure of a Tribal Population, the Yanomama Indians. V. Comparison of a Series of Genetic Networks. Annual Human Genetics, 36, 21-43.

Ward, R. H., Alan Redd, A., Valencia, D., Frazier, B., & Paabo, S. (1993). Genetic and Linguistic Differentiation in the Americas. Proceedings of the National Academy of Sciences, USA, 90, 10663-10667.

Ward, R. H., Frazier, B. L., Dew-Jager, K., & Paabo, S. (1991). Extensive Mitochondrial Diversity within a Single Amerindian Tribe. Proceedings of the National Academy of Sciences, USA, 88, 8720-8724.

Watson, E., Forster, P., Richards, M., & Bandelt, H.-J. (1997). Mitochondrial Footprints of Human Expansions in Africa. American Journal of Human Genetics, 61, 691-704.

Watterson, G. A. (1975). On the Number of Segregating Sites in Genetical Models without Recombination. Theoretical Population Biology, 6, 217-250.

Weiss, K. M. (1994). American Origins. Proceedings of the Academy of Sciences, USA, 91, 833-835.

Wilson, I. J., & Balding, D. J. (1998). Genealogical Inference from Microsatellite Data. Genetics, 150, 499-510.

Wright, S. (1931). Evolution in Mendelian Populations. Genetics, 16, 97-159.

Wright, S. (1943). Isolation by Distance. Genetics, 28, 114-138.

Wright, S. (1951). The Genetical Structure of Populations. Annals of Eugenics, 15, 323-354.

Wright, S. (1969). Evolution and the Genetics of Populations (Vol. 2). Chicago, IL: University of Chicago Press.

Yellen, J., & Harpending, H. (1972). Hunter-Gatherer Populations and Archaeological Inference. World Archaeology, 4, 244-253.

APPENDIX

Study
Bailliet, G., Rothhammer, F., Carnese, F.R., Bravi, C.M., and Bianchi, N.O. 1994. Founder mitochondrial Haplotypes in Amerindian Populations. American Journal of Human Genetics, 55(1): 27-33.

Study Summary
It had been proposed that the colonization of the New World took place by three successive migrations from northeastern Asia. The first one gave rise to Amerindians (Paleoindians), the second and third ones to Na-Dene and Aleut-Eskimo, respectively. Variation in mtDNA has been used to infer the demographic structure of the Amerindian ancestors. The study of RFLP all along the mtDNA and the analysis of nucleotide substitutions in the D-loop region of the mitochondrial genome apparently indicate that most or all full-blooded Amerindians cluster in one of four different mitochondrial haplotypes that are considered to represent the founder maternal lineages of Paleo-Indians. The researchers studied the mtDNA diversity in 109 Amerindians belonging to 3 different tribes, and we have reanalyzed the published data on 482 individuals from 18 other tribes. Their study confirms the existence of four major Amerindian haplotypes. However, they also found evidence supporting the existence of several other potential founder haplotypes or haplotype subsets in addition to the four ancestral lineages reported. Confirmation of a relatively high number of founder haplotypes would indicate that early migration into America was not accompanied by a severe genetic bottleneck.

Genetic Materials and Lines
   Not Available.

Study
   Bergen, A.W., Wang, C.Y., Tsai, J., Jefferson, K., Dey, C., Smith, K.D., et al. 1999. An Asian-Native American Paternal Lineage Identified by RPS4Y Resequencing and by Microsatellite Haplotyping. Annals of Human Genetics, 63: 63-80.

Study Summary
   Human paternal population history was studied in 9 populations (three American Indian, three Asian, two Caucasian and one African-derived sample(s)) using sequence and short tandem repeat haplotype diversity within the non-pseudoautosegmal region of the Y chromosome. Complete coding and additional flanking sequences (949 base pairs) of the RPS4Y locus were determined in 59 individuals from three of the populations, revealing a nucleotide diversity of 0.0147 %, consistent with previous estimates from Y chromosome resequencing studies. One RPS4Y sequence variant, 711C > T, was polymorphic in Asian and Native American populations, but not in African and Caucasian population samples. The RPS4Y 711C > T variant, a second unique sequence variant at DYS287 and nine Y chromosome short tandem repeat (YSTR) loci were used to analyze the evolution of Y chromosome lineages. Three unambiguous lineages were defined in Asian, Native American and Jamaican populations using sequence variants at RPX4Y and DYS287. These lineages were independently supported by the haplotypes defined solely by YSTR alleles, demonstrating the haplotypes constructed from YSTRs can evaluate population diversity, admixture and phylogeny.

Genetic Materials and Lines
   Not Available.

Study

Bianchi, N., Bailliet, G., Bravi, C., Carnese, R., Rothhammer, F., Martinez-Marignac, V. and Pena, S. 1997. Origin of Amerindian Y-Chromosomes as Inferred by the Analysis of Six Polymorphic Markers. American Journal of Physical Anthropology, 102: 79-89.

Study Summary

The researchers analyzed the frequency of six Y-specific polymorphisms in 105 Amerindian males from seven different populations, 42 Caucasian males, and a small number of males of African, Chinese, and Melanesian origin. The combination of three of the six polymorphisms studies produced four different Y-haplogroups. The haplogroup A (non-variant) was the most frequent one. Eighty-five percent of Amerindians showing haplogroup A have the haploid II and the CYS19A Y-specific markers, an association that is found only in 10% of Caucasians and that has not been detected in Asiatics and Africans. Haplogroups C (YAP+) and D (YAP+ plus an A-D transition in the locus DYS271) are of African origin. Four percent of Amerindians and ~12% of Caucasians showed haplogroup C; ~1% of Amerindians and ~2% of Caucasians had haplogroup D. Haplogroup B is characterized by a c-T transition in nucleotide position 373 of the SRY gene domain; this haplogroup is found in Caucasians (~12%) and Amerindians (~4%). None of the Amerindians exhibiting the haplogroups B, C, or D show the haplotype alphahII/DYS19A. By haplotyping the Alu insert and the DNA region surrounding the insert in YAP+ individuals, the researchers could demonstrate that Amerindian Y chromosomes bearing African markers (haplogroups C and D) are due to recent genetic admixture. Most non-alpha hII/DYS19A Amerindian Y-chromosomes in haplogroup A and most cases in haplogroup B are also due to gene flow. The researchers showed that haplotype alpha hII/DYS19A is in linkage disequilibrium with a C-T transition in the locus DYS199. Their results suggest that most Amerindian Y-chromosomes derive from a single paternal lineage characterized by the alpha hII/DYS19A/DYS199T Amerindian-specific haplotype. The analysis of a larger sample of American Indian Y-chromosomes will be required in order to confirm or correct this hypothesis.

Genetic Materials and Lines

The researchers analyzed a total of 105 Amerindian, 42 Caucasian, 3 Chinese, 2 Melanesian, and 4 African pygmy Y-chromosomes. DNA samples from Mapuches, Huilliches, Wichis, Lengua, Tehuelches, Pehuenches, and La Plata Caucasians were obtained from the DNA bank at the IMBICE. All samples from this bank come from donors who gave informed consent to the use of their DNA. CEPH samples were provided by Dr. H.M. Cann from the Centre d'Etude du Polymorphisme Humain, Paris, France. Mayan DNA samples were provided by Dr. R. Herrera from the College of Arts and Sciences, University Park, Miami, Fl.

Study

Bianchi, N.O., Catanesi, C.I., Bailliet, G., Martinez-Marignac, V.L., Bravi, C.M., Vidal-Rioja, L.B., et al. 1998. Characterization of Ancestral and Derived Y-Chromosome Haplotypes of New World Native Populations. American Journal of Human Genetics, 63: 1862-1871.

Study Summary

The researchers analyzed the allelic polymorphisms in seven Y-specific microsatellite loci and a Y-specific alphoid system with 27 variants (alpha h I-XXVII), in a total of 89 Y chromosomes carrying the DYS199T allele and belonging to populations representing Amerindian and Na-Dene linguistic groups. Since there are no indications of recurrence for the DYS199CàT transition, it is assumed that all DYS199T haplotypes derive from a single individual in whom the CàT mutation occurred for the first time. They identified both the ancestral founder haplotype, OA, of the DYS199T lineage and seven derived haplogroups diverging from the ancestral one by one to seven mutational steps. The OA haplotype (5.7% of Native American chromosomes) had the following constitution: DYS199T, alpha h II, DYS19/13, DYS389a/10, DYS389b/27, DYS390/24, DYS391/10, DYS392/14, and DYS393/13 (microsatellite alleles are indicated as number of repeats).

They analyzed the Y-specific microsatellite mutation rate in 1,743 father-son transmissions, and pooled our data with data in the literature, to obtain an average mutation rate of .0012. The researchers estimated that the OA haplotype has an average age-of 22,770 years (minimum 13,500 years, maximum 58,700 years). Since the DYS199T allele is found with high frequency in American Indian chromosomes, they propose that 0A is one of the most prevalent founder paternal lineages of New World aborigines.

Genetic Materials and Lines
Not Available.

Study
Biggar, R. J., Taylor, M. E., Neel, J. V., Hjelle, B., Levine, P. H., Black, F. L., Shaw, G. M., Sharp, P. M. and Hahn, B. H. 1996. Genetic Variants of Human T-Lymphotropic Virus Type II in American Indian Groups. Virology, 216(1): 165-173.

Study Summary
The human T-lymphotropic virus type II (HTLV-II) if found in many New World Indian groups in North and South America and may have entered the New World from Asia with the earliest migration of ancestral Amerindians over 15,000 years ago. To characterize the phylogenetic relationships of HTLV-II strains infecting geographically diverse Indian populations, the researchers used polymerase chain reaction to amplify HTLV-II sequences from lymphocytes of seropositive Amerindians from Brazil (Kraho, Kayapo, and Kaxuyana), Panama (Guaymi), and the United States (the Navajo and Pueblo tribes of the Southwestern states and the Seminoles of Florida). Sequence analysis of a 780-base pair fragment (located between the env gene and the second exons of tax/rex) revealed that Amerindian viruses clustered in the same two genetic subtypes (IIa and IIb) previously identified for viruses from intravenous drug users. Most infected North and Central American Indians had subtype IIb, while HTLV-II infected members of three remote Amazonian tribes clustered as a dis-

tinct group within subtype IIa. These findings suggest that the ancestral Amerindians migrating to the New World brought at least two genetic subtypes, IIa and IIb. Because HTLV-II strains from Amazonian Indians form a distinct group within subtype HTLV-IIa, these Brazilian tribes are unlikely to be the source of IIa viruses in North American drug users. Finally, the near identity of viral sequences from geographically diverse populations indicates that HTLV-II is a very ancient virus of man.

Genetic Materials and Lines

Blood samples were collected during field studies on Amerindians, and cultured HTLV-II isolates derived from Amerindian groups between 1966 and the present. For example, SC, AG, and DSA have been studied here as well as by Ishak et al. (1995), although analyses were done on different genomic regions. For their identification, the researchers include subjects' initials. Both Navajo and Pueblo blood samples were collected in 1990 (Hjelle et al., 1993). Samples were collected in 1989–91 (Levine et al., 1993).

Study

Bonatto, S. and Salzano, F. 1997. Diversity and Age of the Four Major mtDNA Haplogroups, and Their Implications for the Peopling of the New World. American Journal of Human Genetics, 61: 1413-1423.

Study Summary

Despite considerable investigation, two main questions on the origin of American Indians remain the topic of intense debate- namely, the number and time of the migration(s) into the Americas. Using the 720 available Amerindian mtDNA control-region sequences, the researchers reanalyzed the nucleotide diversity found within each of the four major mtDNA haplogroups (A-D) thought to have been present in the colonization of the New World. The researchers first verified whether the within-haplogroup sequence diversity could be used as a measure of the haplogroup's age. The pattern of shared polymorphisms, the mismatch distribution, the phylogenetic trees, the value of Tajima's D, and the computer simulations all suggested that the four haplogroups underwent a bottleneck followed by a large population expansion. The four haplogroup diversities were very similar to each other, offering a strong support for their single origin. They suggested that the beginning of the American Indians' ancestral-population differentiation occurred ~30,000-40,000 years before the present, with a 95% confidence-interval lower bound of ~25,000 ybp. These values are in good agreement with the New World-settlement model that the researchers have presented elsewhere, extending the results initially found for haplogroup A to the three other major groups of mtDNA sequences found in the Americas. These results put the peopling of the Americas clearly in an early, pre-Clovis time frame.

Genetic Materials and Lines

The researchers used all available CR sequences from American Indians were employed, with the exception of the two populations described by Horai et al. (1993), since they were not sequenced for the first 100 bases of HVS-I. For HVS-I, the American Indian sample consists of 720 individuals from a total of 24 populations (with sample sizes n>5) from North, Central, and South America, for each continent, as follows: for South America (n=318) — Xavante (n=25), Zoro (n=30), and Gaviao (n=27) (Ward et al. 1996); Wai Wai (n=30), and Surui (n=24) (researchers' unpublished data); Mapuche (n=39) (Ginther et al. 1993); Yanomama (n=27), Wayampi (n=21), Kayapo (n=13), Arara (n=9), Katuena (n=9), Poturujara (n=9), Awa-Guaja (n=2), and Tiriyo (n=2) (Santos et al. 1996); Yanomami (n=50) (Easton et al. 1996); and Colombian mummies (n=5) (Monsalve et al. 1996); for Central America (n=136)- Huetar (n=27) (Santos et al. 1994); Ngobe (n=46) (Kolman et al. 1995); and Kuna (n=63) (Batista et al. 1995); and, for North America (n=228)- Nuu-Chah-Nulth (n=63) (Ward et al. 1991); Bella Coola (n=40) and Haida (n=41) (Ward et al. 1993); and Yakima (n=42), Athapascan (n=21), Inupiaq Eskimo (n=5), and western Greenland Eskimo (n=16) (Shields et al. 1993). The 38 individuals whose mtDNA Torroni et al. (1993) have sequenced from several populations all over the Americas were also included. For HVS-I+HVS-II, sequences were available from a total of 217 individuals from the Huetar (Santos et al. 1994), Ngobe (Kolman et al. 1995), Mapuche (Ginther et al. 1993), and Yanomami (Easton et al. 1996) and from 24 Surui, 26 Wai Wai, 3 Xavante, 1 Gaviao, and 1 Zoro (authors' unpublished data).

Study

Bonatto, S. L. and Salzano, F. M. 1997. A single and early migration for the peopling of the Americas supported by mitochondrial DNA sequence data. Proceedings of the National Academy of Sciences of the United States of America, 94(5): 1866-1871.

Study Summary

To evaluate the number and time of the migration(s) that colonized the New World the researchers analyzed all available sequences of the first hypervariable segment of the human mitochondrial DNA control region, including 544 American Indians, Sequence and population trees showed that the Amerind, Na-Dene, and Eskimo are significantly closer among themselves than anyone is to Asian populations, with the exception of the Siberian Chukchi, that in some analyses are closer to Na-Dene and Eskimo, Nucleotide diversity analyses based on haplogroup A sequences suggest that American Indians and Chukchi originated from a single migration to Beringia, probably from east Central Asia, that occurred approximate to 30,000 or approximate to 43,000 years ago, depending on which substitution rate is used, with 95% confidence intervals between approximate to 22,000 and approximate to 55,000 years ago. These results support a model for the peopling of the Americas in which Beringia played a central role, where the population that originated the American Indians settled and expanded, Some time after the colonization of Beringia they crossed the Alberta ice-free corridor and peopled the rest of the American continent, The collapse of this ice-free corridor during a few thousand years approximate to 14,000 - 20,000 years ago isolated the people south of the ice-sheets, who gave rise to the Amerind, from those still in

Beringia; the latter originated the Na-Dene, Eskimo, and probably the Siberian Chukchi.

Genetic Materials and Lines

Below are the populations used, for each major geographic region and population group (n = sample size): Amerind: South America (n = 171): Xavante (n = 25), Zoro´ (n = 30), Gaviao (n = 27), Wai-Wai (n = 26), Suru (n = 24), Mapuche (Argentina) (n = 39); Central America (n = 136): Huetar (n = 27), Ngobe´ (n = 46), Kuna (n = 63); North America (n = 145): Nuu-Chah-Nulth (n = 63), Yakima (n = 42), Bella Coola (n = 40); Na-Dene (n = 70): Haida (n = 41), Athapaskan (n = 21), several populations (n = 8); Eskimo (n = 22): Inupiaq Eskimo (n = 5), West Greenland Eskimo (n = 17); Siberian Chukchi (n = 7); Siberia (n = 33): Altai (n = 17), several populations (n = 16); Mongolia (n = 103); East Asia (n = 99); Africa (n = 42): Mbuty Pygmy (n = 30), Yoruba (n = 12); Europe (n = 77).

Study

Bortolini, M. C., Salzano, F. M., Bau, C. H., Layrisse, Z., Petzl-Erler, M. L., Tsuneto, L. T., Hill, K., Hurtado, A. M., Castro-De-Guerra, D., Bedoya, G. and Ruiz-Linares, A. 2002. Y-chromosome biallelic polymorphisms and American Indian population structure. Annual Review of Human Genetics, 66(Pt 4): 255-9.

Study Summary

It has been proposed that women had a higher migration rate than men throughout human evolutionary history. However, in a recent study of South American natives using mtDNA restriction fragment polymorphisms and Y-chromosome microsatellites the researchers failed to detect a significant difference in estimates of migration rates between the sexes. As the high mutation rate of microsatellites might affect estimates of population structure, the researchers now examined biallelic polymorphisms in both mtDNA and the Y-chromosome. Analyses of these markers in Amerinds from North, Central and South America agree with the researchers previous findings in not supporting a higher migration rate for women in these populations. Furthermore, they underline the importance of genetic drift in the evolution of Amerinds and suggest the existence of a North to South gradient of increasing drift in the Americas.

Genetic Materials and Lines

The Brazilian Fundacao Nacional do Indio authorized this study. The subjects of this investigation were informed about the aims of the study and gave their consent. Some of the samples examined were collected many years ago as part of a long-term association that F. M. Salzano had with the late Prof. James V. Neel and with Prof. Francis L. Black. The Brazilian National Ethics Commission approved this investigation (CONEP Resolution no.123/98).

Y-chromosome typings were performed in 356 unrelated male samples from 17 Native South American populations: 8 from Brazil, 5 from Colombia, 3 from Venezuela and one from Paraguay. This population sample includes representatives of the four major linguistic subfamilies present in South America Andean, Chibchan-Paezan, Equatorial-Tucano and Ge-Pano-Carib.

Study

Bortolini, M. C., Salzano, F. M., Thomas, M. G., Stuart, S., Nasanen, S. P., Bau, C. H., Hutz, M. H., Layrisse, Z., Petzl-Erler, M. L., Tsuneto, L. T., Hill, K., Hurtado, A. M., Castro-de-Guerra, D., Torres, M. M., Groot, H., Michalski, R., Nymadawa, P., Bedoya, G., Bradman, N., Labuda, D. and Ruiz-Linares, A. 2003. Y-chromosome evidence for differing ancient demographic histories in the Americas. American Journal of Human Genetics, 73(3): 524-39.

Study Summary

To scrutinize the male ancestry of extant American Indian populations, the researchers examined eight biallelic and six microsatellite polymorphisms from the nonrecombining portion of the Y chromosome, in 438 individuals from 24 American Indian populations (1 Na-Dene and 23 South

Amerinds) and in 404 Mongolians. One of the biallelic markers typed is a recently identified mutation (M242) characterizing a novel founder American Indian haplogroup. The distribution, relatedness, and diversity of Y lineages in American Indians indicate a differentiated male ancestry for populations from North and South America, strongly supporting a diverse demographic history for populations from these areas. These data are consistent with the occurrence of two major male migrations from southern/central Siberia to the Americas (with the second migration being restricted to North America) and a shared ancestry in central Asia for some of the initial migrants to Europe and the Americas. The microsatellite diversity and distribution of a Y lineage specific to South America (Q-M19) indicates that certain Amerind populations have been isolated since the initial colonization of the region, suggesting an early onset for tribalization of American Indians. Age estimates based on Y-chromosome microsatellite diversity place the initial settlement of the American continent at approximately 14,000 years ago, in relative agreement with the age of well-established archaeological evidence.

Genetic Materials and Lines

Researchers used samples that were collected in collaboration with Professors F. M. Salzano, Francis L. Black, and the late James V. Neel. This investigation was approved by the Brazilian National Ethics Commission, the Canadian Institutional Review Boards of the Sainte-Justine (Montreal) and Victoria (Prince Albert) Hospitals, The Prince Albert Grand Council, and the Bioethics Committee of Universidad de Antioquia (Colombia).

The total number of unrelated American Indian males available for typing was 438 from 24 populations. The Na Dene linguistic family is represented by a broad sample of Chipewayan speakers (n = 48) from several locations in Saskatchewan, Canada (primarily Fond-du-Lac and Uranium City). The South American populations were sampled in Brazil, Colombia, Paraguay, and Venezuela and include representatives of the four main subdivisions of South Amerind: Andean (Ingano, n=9), Chibchan-Paezan (Barira, n=12; and Warao, n=12), Equatorial Tucano (Ache, n=54; Asurini, n=4;

Cinta-Larga, n=15; Guarani, n=59; Paacas Novos, n=15; Parakana, n=20; Ticuna, n=33; Urubu-Kaapor, n=16; Waiapi, n=14; and Wayuu, n=19), and Ge-Pano-Carib (Gorotire, n=5; Huitoto, n=4; Kaigang, n=22; Kraho, n=9; Mekranoti, n=7; Tiryio , n=4; Xikrin, n=8; Yagua, n=7; and Yukpa, n=12). The linguistic affiliation of one population (Zenu, n=30) is undefined. The Mongolian sample consisted mostly of Khalkh (n=303), Buryad (n=23), and Durvud (n=29), as well as a smaller number of individuals from 11 other ethnic groups (Barga, n=4; Bayad, n=10; Dariganga, n=1; Darkhad, n=4; Kazakh, n=6; Myangad, n=3; Torguud, n=3; Uryankhai, n=4; Uuld, n=2; Uzemchim, n=4; and Zakhchin, n=8). For most analyses, the 404 Mongolian individuals were considered as a single sample.

Study

Bravi, C. M., Bailliet, G., Martinez-Marignac, V. L. and Bianchi, N. O. 2001. Tracing the origin and geographic distribution of an ancestral form of the modern human Y chromosome. Revista Chilena De Historia Natural, 74(1): 139-149.

Study Summary

The researchers screened a total of 841 Y chromosomes representing 36 human populations of wide geographical distribution for the presence of a Y-specific Alu insert (YAP+ chromosomes). The Alu element was found in 77 cases. They tested five biallelic and eight polyallelic markers in 70 out of the 77 YAP+ chromosomes. They could identify the existence of a hierarchical and chronological structuring of ancestral and derived YAP+ lineages giving rise to four haplogroups, 14 subhaplogroups and 60 haplotypes. Moreover, the researchers proposed a monophyletic origin for each one of the YAP+ lineages. Out-of-Africa and out-of-Asia models have been suggested to explain the origin and evolution of ancestral and derived YAP+ elements. They analyzed the evidence supporting these two hypotheses and the researchers conclude that the information available supports better the out-of-Africa model.

Genetic Materials and Lines

The researchers studied a total of 841 Y chromosomes representing different populations and geographic regions. Allele DYS199*T is specific of American Indians; 64 % of the Amerindian Y chromosomes in the researchers series showed this marker. In this report, all cases of YAP+ chromosomes in American Indians had the ancestral allele DYS199*C indicating that these chromosomes are the result of European (haplogroup C) or African (haplogroup D) admixture. Samples were provided by M. Hammer (one Tibetan and one Japanese, one Bantu and two Gambians), Y-F Chris Lau (20 Chinese and all Laotians, Cambodians, Thai, Vietnamese and Philippines), M. Sans (Afro-Uruguayans), G. Cantos (Afro-Ecuadorians), R. Herrera (Afro-USA, Pakistanis and Bengalis, Chimila, Maya, Zuñi, Sioux, Navajo), J. Ferrer (Ayoreo and Lengua), P. Zukas (Mocovies), F. Rothhammer (Huilliche and Pehuenche), F. Carnese (Mapuche, Tehuelche, Wichi, Chorote and Toba), Nippon University Association of La Plata (12 Japanese), Lebanese Society of La Plata and Argentine Arab Home of Berisso (Lebanese), Syrian-Orthodox Association of La Plata (Syrians), Centre d Etude du Polymorphismes Humains (CEPH pedigrees), Coriell (Pygmies and Melanesian) IMBICE DNA repository (La Plata and Jews). As far as the researchers know CEPH samples were not tested before for YAP+ chromosomes. Therefore, 829 out of the 841 Y chromosomes tested for Alu inserts and the information on the ah system and microsatellites are new data produced by the researchers group.

Study

Brown, M. D., Hosseini, S. H., Torroni, A., Bandelt, H.-J., Allen, J. C., Schurr, T. G., Scozzari, R., Cruciani, F. and Wallace, D. C. 1998. mtDNA Haplogroup X: An Ancient Link between Europe/Western Asia and North America. American Journal of Human Genetics, 63: 1852-1861.

Study Summary

On the basis of comprehensive RFLP analysis, it has been inferred that ~97% of American Indian mtDNAs belong to one of the researchers major founding mtDNA lineages, designated haplogroups A-D. It has been proposed that a fifth mtDNA haplogroup (haplogroup X) represents a minor founding lineage in American Indians. Unlike haplogroups A-D, haplogroup X is also found at low frequencies in modern European populations. To investigate the origins, diversity, and continental relationships of this haplogroup, the researchers preformed mtDNA high-resolution RFLP and complete control region (CR) sequence analysis on 22 putative American Indian haplogroup X and 14 putative European haplogroup X mtDNAs. The results identified a consensus haplogroup X motif that characterizes the researchers European and American Indian samples. Among American Indians, haplogroup X appears to be essentially restricted to northern Amerindian groups, including the Ojibwa, the Nuu-Chah-Nulth, the Sioux, and the Yakima, although they also observed this haplogroup in the Na-Dene speaking Navajo. Median network analysis indicated that European and American Indian haplogroup X mtDNAs, although distinct, nevertheless are distantly related to each other. Time estimates for the arrival of X in North America are 12,000-36,000 years ago, depending on the number of assumed founders, thus supporting the conclusion that the peoples harboring haplogroup X were among the original founders of American Indian populations. To date, haplogroup X has not been unambiguously identified in Asia, raising the possibility that some American Indian founders were of Caucasian ancestry.

Genetic Materials and Lines

The researchers obtained samples from the following individuals: Drs. R. Herrera, K. Kidd, J. Kidd, B. Bonne-Tamir, D. Cole, D. Labuda, R. Fourney, C. Fregeau, G. Troup, D. Smith, M.-L. Savontaus, and R. Sukernik.

A total of 36 (22 American Indian and 14 European) individuals were available in the researchers collection, as putative haplogroup X samples, and were used for high-resolution RFLP and CR sequence analysis. The 22 American Indians (designated "NA1" - "NA22") consisted of 7 northern Ojibwa (NA1, NA2, NA7-NA10, and NA20) from the northwestern region of Ontario, 2 southwestern Ojibwa samples

(NA3 and NA21) from Wisconsin, 5 southeastern Ojibwa (NA4 - NA6, NA11, and NA22) from Manitoulin Island in Lake Huron, 2 Navajo (NA14 and NA15) from New Mexico, 2 Nuu-Chah-Nulth (NA12 and NA13) from Vancouver Island, British Columbia, and 4 Navajo (NA16-NA19) collected from New Mexico by Dr. Rene Herrera. The 14 Caucasian-European haplogroup X samples (designated "CE1" - "CE14") included 2 Caucasians of European ancestry (CE1 and CE4) from the United States and 1 French Canadian (CE5), (Torroni et al. 1994a), 1 Finn (CE8), 5 Israeli Druze (CE2 and CE11-CE14), and 5 Italians (CE3, CE6, CE7, CE9, and CE10).

Study

Calleja-Macias, I. E., Kalantari, M., Huh, J., Ortiz-Lopez, R., Rojas-Martinez, A., Gonzalez-Guerrero, J. F., Williamson, A.-L., Hagmar, B., Wiley, D. J., Villarreal, L., Bernard, H.-U. and Barrera-Saldana, H. A. 2004. Genomic Diversity of Human Papillomavirus-16, 18, 31, and 35 Isolates in a Mexican Population and Relationship to European, African, and American Indian Variants. Virology, 319: 315-323.

Study Summary

Cervical cancer, mainly caused by infection with human papillomaviruses (HPVs), is a major public health problem in Mexico. During a study of the prevalence of HPV types in northeastern Mexico, the researchers identified, as expected from worldwide comparisons, HPV-16, 18, 31, and 35 as highly prevalent. It is well known that the genomes of HPV types differ geographically because of evolution linked to ethnic groups separated in prehistoric times. As HPV intra-type variation results in pathogenic differences, they analyzed genomic sequences of Mexican variants of these four HPV types. Among 112 HPV-16 samples, 14 contained European and 98 American Indian (AA) variants. This ratio is unexpected as people of European ethnicity predominate in this part of Mexico. Among 15 HPV-18 samples, 13 contained European and 2 African variants, the latter possibly due to migration of Africans to the Caribbean coast of Mexico. They constructed phylogenetic trees of HPV-31 and 35 variants, which have never been studied. Forty-six HPV-31 isolates from Mexico, Europe, Africa, and the United States (US) contained a total of 35 nucleotide exchanges in a 428-bp segment, with maximal distances between any two variants of 16 bp (3.7%), similar to those between HPV-16 variants. The HPV-31 variants formed two branches, one apparently the European, the other one an African branch. The European branch contained 13 of 29 Mexican isolates, the African branch 16 Mexican isolates. These may represent the HPV-31 variants of American Indians, as a 55% prevalence of African variants in Mexico seems incomprehensible. Twenty-seven HPV-35 samples from Mexico, Europe, Africa, and the US contained 11 mutations in a 893-bp segment with maximal distances between any two variants of only 5 mutations (0.6%), including a characteristic 16-bp insertion/deletion. These HPV-35 variants formed several phylogenetic clusters rather than two- or three-branched trees as HPV-16, 18, and 31. An HPV-35 variant typical for American Indians was not identifiable. The researchers suggested type specific patterns of evolution and spread of HPV-16, 18, 31, and 35 both before and after the worldwide migrations of the last four centuries. The high prevalence of highly carcinogenic HPV-16 AA variants, and the extensive diversity of HPV-18, 31, and 35 variants with unknown pathogenic properties raise the possibility that HPV intra-type variation contributes to the high cervical cancer burden in Mexico.

Genetic Materials and Lines

Among a cohort of 1200 consecutive women from several primary health care centers in Monterrey, Nuevo Leon, Mexico, a total of 161 specimens were positive for the four most common HPV types, namely 112 for HPV-16, 29 for HPV-31, 20 for HPV-18, and 7 for HPV-35 positive samples. All of these samples entered this study (designated as MX-Z, Z being the code number of the patients), except five samples with HPV-18 with insufficient DNA. These samples had been taken in the form of swabs during a large ongoing epidemiological study. The 13 Norwegian swabs (NW-Z) were collected in Oslo, Norway, during gynecological consultations of these patients. Sixteen South Africa specimens (swabs) from

Cape Town excluded white patients but were from two different ethnic groups that the researchers refer to as black (SA*-Z) and mixed race (SA-Z) women. For all three cohorts, cytological diagnoses were done after obtaining these samples, and the clinical outcome did not influence the inclusion in this study. The samples from the United States of America were obtained in Los Angeles and derived from biopsies of anal neoplastic lesions in patients positive for infection with the human immunodeficiency virus. The use of these samples was approved by the Institutional Review Board of the University of California, Irvine, and collection followed the respective patient protection rules of each of the four participating clinics in Monterrey, Oslo, Cape Town, and Los Angeles.

Study

Carlyle, S.W., Parr, R.L., Hayes, M.G., and O'Rourke, D.H. 2000. Context of Maternal Lineages in the Greater Southwest. American Journal of Physical Anthropology, 113(1): 85-101.

Study Summary

The researchers present mitochondrial haplogroup characterizations of the prehistoric Anasazi of the United States (US) Southwest. These data are part of a long-term project to characterize ancient Great Basin and US Southwest samples for mitochondrial DNA (mtDNA) diversity. Three restriction site polymorphisms (RSPs) and one length polymorphism identify four common American Indian matrilines (A, B, C, and D). The Anasazi (n = 27) are shown to have a moderate frequency of haplogroup A (22%), a high frequency of haplogroup B (56%), and a low frequency of C (15%). Haplogroup D has not yet been detected among the Anasazi, In comparison to modern American Indian groups from the US Southwest, the Anasazi are shown to have a distribution of haplogroups similar to the frequency pattern exhibited by modern Pueblo groups. A principal component analysis also clusters the Anasazi with some modern (Pueblo) Southwestern populations, and away from other modern (Athapaskan speaking) Southwestern populations. The Anasazi are also shown to have a significantly different distribution of the four haplogroups as compared to the eastern Great Basin Great Salt Lake Fremont (n = 32), although both groups cluster together in a principal component analysis. The context of our data suggests substantial stability within the US Southwest, even in the face of the serious cultural and biological disruption caused by colonization of the region by European settlers. We conclude that although sample numbers are fairly low, ancient DNA (aDNA) data are useful for assessing long-term populational affinities and for discerning regional population structure.

Genetic Materials and Lines
Not Available.

Study
Cerda-Flores, R. M., Budowle, B., Jin, L., Barton, S. A., Deka, R. and Chakraborty, R. 2002. Maximum Likelihood Estimates of Admixture in Northeastern Mexico Using 13 Short Tandem Repeat Loci. American Journal of Human Biology, 14: 429-439.

Study Summary
Tetrameric short tandem repeat (STR) polymorphisms are widely used in population genetics, molecular evolution, gene mapping and linkage analysis, paternity tests, forensic analysis, and medical applications. This research provided allelic distributions of the STR loci D3S1358, vWA, FGA, D8S1179, D21S11, D18S51, D5S818, D13S317, D7S820, CSF1PO, TPOX, TH01, and D16S539 in 143 Mestizos from Northeastern Mexico, estimates of contributions of genes of European (Spanish), American Indian and Africa origin in the gene pool of this admixed Mestizo population (using 10 of these loci); and a comparison of the genetic admixture of this population with the previously reported two polymorphic molecular markers, D1S80 and HLA-DQA1 (n=103). Genotype distributions were in agreement with Hardy-Weinberg expectations (HWE) for almost all 13 STR markers. Maximum likelihood estimates of admixture components yield a trihybrid model with Spanish, Amerindian, and

African ancestry with the admixture proportions: 54.99% +/- 3.44, 39.99% +/- 2.57, and 5.02% +/-2.82, respectively. These estimates were not significantly different from those obtained using D1S80 and HLA-DQA1 loci (59.99% +/- 5.94, 36.99% +/- 5.04, and 3.02% +/-2.76). In conclusion, Mestizos of Northeastern Mexico showed a similar ancestral contribution independent of the markers used for evolutionary purposes. Further validation of this database supports the use of the 13 STR loci along with D1S80 and HLA-DQA1 as a battery of efficient DNA forensic markers in Northeastern Mestizo populations of Mexico

Genetic Materials and Lines

The data from this population were collected as part of a larger investigation of the genetic structure of the Mexican Mestizo populations in Northeastern Mexico. Whole blood (5 ml) was collected into tubes containing EDTA by venipuncture from unrelated, healthy individuals. 143 unrelated Mestizo individuals were sampled from the greater Monterrey Metropolitan Area of Nuevo Leon, Mexico.

Study

Cerda-Flores, R. M., Villalobos-Torres, M. C., Barrera-Saldana, H. A., Cortes-Prieto, L. M., Barajas, L. O., Rivas, F., Carracedo, A., Zhong, Y., Barton, S. A. and Chakraborty, R. 2002. Genetic Admixture in Three Mexican Mestizo Populations Based on D1S80 and HLA-DQA1 Loci. American Journal of Human Biology, 14: 257-263.

Study Summary

This study compares genetic polymorphisms at the D1S80 and HLA-DQA1 loci in three Mexican Mestizo populations from three large states (Nuevo Leon, Jalisco, and the Federal District). Allele frequency distributions are relatively homogenous in the three samples; only the Federal District population shows minor differences of the HLA-DQA1 allele frequencies compared with the other two. In terms of genetic composition, these Mestizo populations show evidence of admixture with predominantly Spanish-European (50–60%)

and Amerindian (37–49%) contributions; the African contribution (1–3%) is minor. Together with the observation that in Nuevo Leon, the admixture estimates based on D1S80 and HLA-DQA1, are virtually the same as those reported earlier from blood group loci, suggests that DNA markers, such as D1S80 and HLA-DQA1 are useful for examining genetic homogeneity/heterogeneity across Mestizo populations of Mexico. The inverse relationship of the proportion of gene diversity due to population differences (Gst) to within population gene diversity (Hs) is also consistent with theoretical predictions, supporting the use of these markers for population genetics studies.

Genetic Materials and Lines

Genetic data from these populations were collected as part of a larger investigation of the genetic structure of the Mexican Mestizo populations. Venous blood samples from unrelated healthy individuals were collected in tubes containing EDTA. The Nuevo Leon sample consists of 103 individuals, interviewed at the Universidad Autonoma de Nuevo Leon (74 students) and in the Instituto Mexicano del Seguro Social (29 white-collar workers) from 1997 to 1998. In the Jalisco sample, 129 individuals who were scored for D1S80 and 63 for HLADQA1, were unrelated individuals living in the Guadalajara metropolitan area. All of them had Mexican Mestizo parents and grandparents, mainly from the state of Jalisco. The sampled individuals were either family members of Instituto Mexicano del Seguro Social employees or university student volunteers. The sample from the Federal District, chosen for comparative purposes, consists of allele frequency data for D1S80 and HLA-DQA1, obtained from the compilation of Peterson et al. (2000).

Study

Chakraborty, R., Stivers, D. N., Su, B., Zhong, Y. and Budowle, B. 1999. The utility of short tandem repeat loci beyond human identification: implications for development of new DNA typing systems. Electrophoresis, 20(8): 1682-1696.

Study Summary

Since the first characterization of the population genetic properties of repeat polymorphisms, the number of short tandem repeat (STR) loci validated for forensic use has now grown to at least 13. Worldwide variations of allele frequencies at these loci have been studied, showing that variations of interpopulation diversity at these loci do not compromise the power of identification of individuals. However, data collected for validation of these loci for forensic use has utility beyond human identification; the origin and past migration history of modern humans can be reconstructed from worldwide variations at these loci. Furthermore, complex forensic cases previously irresolvable can now be investigated with the help of the validated STR loci. Here, the researchers provide the absolute power of the validated set of 13 STR loci for addressing these issues using multilocus genotype data on 1,401 individuals belonging to seven populations (US European-American, US African-American, Jamaican, Italian, Swiss, Chinese and Apache Native-American). Genomic research is discovering new classes of polymorphic loci (such as the single nucleotide polymorphisms, SNPs) and lineage markers (such as the mitochondrial DNA and Y-chromosome markers); their aim, therefore, was to determine how many SNP loci are needed to match the power of this set of 13 STR loci. They conclude that the current set of STR loci is adequate for addressing most problems of human identification (including interpretations of DNA mixtures). However, if suitable number of SNPs are used that would match the power of the STR loci, they alone cannot resolve more complex cases unless they are supplemented by the validated STR loci.

Genetic Materials and Lines

In this presentation, the researchers considered allele frequency data at the 13 STR loci, which constitute a commonly used battery of forensic loci, in several representative world populations, e.g., Chinese, US European-American, Swiss, Italian, US Native-American (Apache), US African-American, and Jamaican. Estimates of allele frequencies are obtained from multilocus genotype data in 117+/-244 individuals from each population, selected for the study without any prior knowledge of their DNA type.

Study

Derenko, M. V., Malyarchuk, B. A. and Dambueva, I. K. 2000. Mitochondrial DNA variation in two South Siberian aboriginal populations: implications for genetic history of North Asia. Human Biology, 72(6): 945-73.

Study Summary

The mtDNAs of 76 individuals representing the aboriginal populations of South Siberia, the Tuvinians and Buryats, were subjected to restriction fragment length polymorphism (RFLP) analysis and control region hypervariable segment I (HVS-I) sequencing, and the resulting data were combined with those available for other Siberian and East Asian populations and subjected to statistical and phylogenetic analysis. This analysis showed that the majority of the Tuvinian and Buryat mtDNAs (94.4 percent and 92.5 percent, respectively) belong to haplogroups A, B, C, D, E, F, and M*, which are characteristic of Mongoloid populations. Furthermore, the Tuvinians and Buryats harbor four Asian- and American Indian-specific haplogroups (A-D) with frequencies (72.2 percent and 55 percent, respectively) exceeding those reported previously for Mongolians, Chinese, and Tibetans. They represent, therefore, the populations that are most closely related to New World indigenous groups. Despite their geographical proximity, the Tuvinians and Buryats shared no HVS-I sequences in common, although individually they shared such sequences with a variety of other Siberian and East Asian populations. In addition, phylogenetic and principal component analyses data of mtDNA sequences show that the Tuvinians clustered more closely with Turkic-speaking Yakuts, whereas the Mongolic-speaking Buryats clustered closer to Korean populations. Furthermore, HVS-I sequences, comprising one-fourth of the Buryat lineages and characterized by the only C-to-T transition at nucleotide position 16223, were identified as different RFLP haplotypes (B, C, D, E, M*, and H). This finding appears to indicate the putative ancestral state of the 16223T HVS-I sequences to Mongoloid macrohaplogroup

M, at least. Finally, the results of nucleotide diversity analysis in East Asian and Siberian populations suggest that Central and East Asia were the source areas from which the genetically heterogeneous Tuvinians and Buryats first emerged.

Genetic Materials and Lines

In this study, the researchers first performed an analysis of mtDNA variation in two South Siberian aboriginal populations: Tuvinians and Buryats. They reported and analyzed combined RFLP haplotypes and HVS-I sequences of 76 subjects from two adjacent geographic areas (Lake Baikal region and the Altai and Sayan Mountains) that are thought to be close to the origins of important population expansions into Eurasia, to present a detailed picture of the origins of their mtDNA gene pool. The populations studied encompass a large portion of the linguistic and genetic differentiation of modern Siberian populations.

DNA was extracted from the hair roots of 40 Buryats and 36 Tuvinians. All individuals were unrelated, and all stated that their maternal grandmother had been born in the area considered for this study. Samples from Tuvinians were collected in various villages of western Tuva regions including Barun-Khemchiksk, Dzun-Khemchiksk, Bai-Taiga, and Sut-Kholsk districts. The Buryats were sampled in the villages of the Buryat Republic (Dzhida, Eravna, Bichursk, Zakamensk, Ivolga, Kabansk, Kizhinga, Kyakhta, Muhorshibirsk, Selenga, Tunka, and Khorinsk districts), Chita (Aginsk district), and Irkutsk (Alarsk, Bayandaevsk, Bokhan, Nukutsk, Olkhon, Ust-Orda, Osa, Ekhirit Balagansk districts), regions encompassing all territories inhabited by modern Buryats. Only a few samples were chosen from each locality, limiting the probability of sampling relatives.

Study

Easton, R. D., Merriwether, A., Crews, D. E. and Ferrell, R. E. 1996. mtDNA Variation in the Yanomami: Evidence for Additional New World Founding Lineages. American Journal of Human Genetics, 59: 213-225.

Study Summary

American Indians have been classified into four founding haplogroups with as many as seven founding lineages based on mtDNA RFLPs and DNA sequence data. mtDNA analysis was completed for 83 Yanomami from eight villages in the Surucucu and Catrimani Plateau regions of Roraima in northwestern Brazil. Samples were typed for 15 polymorphic mtDNA sites (14 RFLP sites and 1 deletion site), and a subset was sequenced for both hypervariable regions of the mitochondrial D-loop. Substantial mitochondrial diversity was detected among the Yanomami, five of seven accepted founding haplotypes and three others were observed. Of the 83 samples, 4 (4.8%) were lineage B1, 1 (1.2%) was lineage B2, 31 (37.4%) were lineage C1, 29 (34.9%) were lineage C2, 2 (2.4%) were lineage D1, 6 (7.2%) were lineage D2, 7 (8.4%) were a haplotype the researchers designated "X6," and 3 (3.6%) were a haplotype they designated "X7." Sequence analysis found 43 haplotypes in 50 samples. B2, X6, and X7 are previously unrecognized mitochondrial founding lineage types of American Indians. The widespread distribution of these haplotypes in the New World and Asia provides support for declaring these lineages to be New World founding types.

Genetic Materials and Lines

Dr. J.J. Mancilha-Carvalho supplied the original samples for this study. During January and February, 1990, Dr. Jairo J. Mancilha-Carvalho and a Brazilian Ministry of Health team visited eight villages of the Surucucu and Catrimani Plateau regions of the Territorio de Roraima in northwestern Brazil. Blood was drawn from an average of 12 to 13 adult volunteers in each village, giving a total of 61 men and 39 women. Most of these individuals were not first-degree relatives, although more distantly related individuals were included. No pedigree or village-of-origin data were obtained as part of this survey.

Study
Eshleman, J. A., Malhi, R. S., Johnson, J. R., Kaestle, F., Lorenz, J. and Smith, D. G. 2004. Mitochondrial DNA and Prehistoric Settlements: native Migrations on the Western Edge of North America. Human Biology, 76(1): 55-75.

Study Summary
The researchers analyzed previously reported mtDNA haplogroup frequencies of 577 individuals and hypervariable segment 1 (HVS1) sequences of 265 individuals from American Indian tribes in western North America to test hypotheses regarding the settlement of this region. These data were analyzed to determine whether Hokan and Penutian, two hypothesized ancient linguistic stocks, represent biological units as a result of shared ancestry within these respective groups. Although the pattern of mtDNA variation suggests regional continuity and although gene flow between populations has contributed much to the genetic landscape of western North America, some evidence supports the existence of both the Hokan and Penutian phyla. In addition, a comparison between coastal and inland populations along the west coast of North America suggests an ancient coastal migration to the New World. Similarly high levels of haplogroup A among coastal populations in the Northwest and along the California coast as well as shared HVS1 sequences indicate that early migrants to the New World settled along the coast with little gene flow into the interior valleys.

Genetic Materials and Lines
Haplogroup identities of 584 American Indians, representing 11 populations, and sequences from a segment of the first hypervariable region (HVS1) of the control region (np 16,090–16365) of 265 of these individuals. Sequences that could not be assigned to one of the five known American Indian founding haplogroups were not included in these analyses. Although some unassigned sequences might represent yet undocumented founding American Indian mtDNA types, the frequency of "other" types is rather low and likely represents recent admixture.

Study

Falush, D., Wirth, T., Linz, B., Pritchard, J. K., Stephens, M., Kidd, M., Blaser, M. J., Graham, D. Y., Vacher, S., Perez-Perez, G. I., Yamaoka, Y., Megraud, F., Otto, K., Reichard, U., Katzowitsch, E., Wang, X., Achtman, M. and Suerbaum, S. 2003. Traces of Human Migrations in Helicobacter pylori Populations. Science, 299: 1582-1585.

Study Summary

Helicobacter pylori, a chronic gastric pathogen of human beings, can be divided into seven populations and subpopulations with distinct geographical distributions. These modern populations derive their gene pools from ancestral populations that arose in Africa, Central Asia, and East Asia. Subsequent spread can be attributed to human migratory uses such as the prehistoric colonization of Polynesia and the Americas, the neolithic introduction of farming to Europe, the Bantu expansion within Africa, and the slave trade.

Genetic Materials and Lines

The researchers sequenced the same fragments from 370 strains isolated from 27 geographical, ethnic, and/or linguistic human groupings. Of the 3850 nucleotides sequenced for each isolate, 1418 were polymorphic and were used to define bacterial populations.

Study

Forster, P., Harding, R., Torroni, A. and Bandelt, H.-J. 1996. Origin and Evolution of American Indian mtDNA Variation: A Reappraisal. American Journal of Human Genetics, 59: 935-945.

Study Summary

The timing and number of prehistoric migrations involved in the settlement of the American continent is subject to intense debate. Here, the researchers reanalyzed American Indian control region mtDNA data and demonstrated that only an appropriate phylogenetic analysis accompanied by an appreciation of demographic factors allows one to discern different migrations and to estimate their ages. Reappraising 574 mtDNA control region sequences from aboriginal Siberians and American Indians, the researchers confirm in agreement with linguistic, archaeological, and climatic evidence that 1) the major wave of migration brought one population, ancestral to the Amerinds, from northeastern Siberia to America 20,000-25,000 ya, and 2) a rapid expansion of a Beringian source population took place at the end of the Younger Dryas glacial phase ~11,300 ya, ancestral to present Eskimo and Na-Dene populations.

Genetic Materials and Lines

Boris Malyarchuk (Magadan) provided access to the Siberian data. American Indian and Siberian sequences were taken from the following publications: 1) Ward et al. (1991), 2) Ward et al. (1993), 3) Shields et al. (1993), 4) Ivanova (1993) and Voevoda et al. (1994), 5) Torroni et al. (1993a, 1993b), 6) Ginther et al. (1993), 7) Horai et al. (1993), 8) Santos et al. (1994), 9) Batista et al. (1995), and 10) Kolman et al. (1995).

Study

Hammer, M. F. and Zegura, S. L. 2002. The Human Y Chromosome Haplogroup Tree: Nomenclature and Phylogeography of Its Major Divisions. Annual Review of Anthropology, 31(1): 303-321.

Study Summary

In this review the researchers discuss the recent construction of a highly resolved tree of the nonrecombining portion of the Y chromosome (NRY), and the development of a cladistic nomenclatural system to name the resulting haplogroups. This phylogenetic gene tree comprises 18 major haplogroups that are defined by 48 binary polymorphisms. The researchers also present results from a phylogeographic analysis of NRY haplogroups in a global sample of 2007 males, as well as from a regional study focusing on Siberia (n = 902). The researchers use the following statistical techniques to explicate the researchers presentation: analysis of molecular variance, multidimensional scaling, comparative measures of genetic diversity, and phylogeography-based frequency distributions. Their global results, based on the 18 major haplogroups, are similar to those from previous analyses employing additional markers and support the hypothesis of an African origin of human NRY diversity. Although Africa exhibits greater divergence among haplogroups, Asia contains the largest number of major haplogroups (N = 15). The multidimensional scaling analysis plot indicates that the Americas, Africa, and East Asia are outliers, whereas the rest of the world forms a large central cluster. According to the researchers new global-level analysis of molecular variance, 43% of the total variance of NRY haplogroups is attributable to differences among populations (i.e., PhiST = 0.43). The Siberian regional analysis of 62 binary markers exhibits nonrandom associations between geographically restricted NRY haplogroups and language families. The researchers conclude with a list of typing recommendations for laboratories that wish to use the Y chromosome as a tool to investigate questions of anthropological interest.

Genetic Materials and Lines
Not Available.

Study
Horai, S., Kondo, R., Nakagawa-Hattori, Y., Hayashi, S., Sonoda, S. and Tajima, K. 1993. Peopling of the Americas, founded by four major lineages of mitochondrial DNA. Molecular Biological Evolution, 10(1): 23-47.

Study Summary
Nucleotide sequence analysis of the major noncoding region of human mitochondrial DNA from various races was extended with 72 American Indians from 16 different local populations (nine populations from Chile, four from Colombia, and one each from Brazil and from Maya and Apache Indians). The sequences were determined directly from the polymerase chain reaction products. On the basis of a comparison of the 482-bp sequences in the 72 American

Indians, 43 different types of mitochondrial DNA sequences were observed. The nucleotide diversity within the American Indians was estimated to be 1.29%, which is slightly less than the value of 1.44% from the total human population including Africans, Europeans, and Asians. Phylogenetic analysis revealed that most American Indian lineages are classified into four major distinct clusters. Individuals belonging to each cluster share at least two specific polymorphic sites that are nearly absent in other human populations, indicating a unique phylogenetic position of American Indians. A phylogenetic tree of 193 individuals including Africans, Europeans, Asians, and American Indians indicated that the four American Indian clusters are distinct and dispersed in the tree. These clusters almost exclusively consist of American Indians — with only a few Asians, if any. The researchers postulate that four ancestral populations gave rise to different waves of migration to the New World. From the estimated coalescence time of the Asian and American Indian lineages, they infer that the first migration across the Bering landbridge took place approximately 14,000-21,000 years ago. Furthermore, sequence differences in all pairwise comparisons of American Indians showed a bimodal distribution that is significantly different from Poisson. These results suggest that the ancestral American Indian population underwent neither a severe bottleneck nor rapid expansion in population size, during the migration of people into the Americas.

Genetic Materials and Lines

Blood lymphocytes were obtained from 65 subjects including 20 individuals from four locations in Colombia and 45 individuals from nine locations in Chile who were randomly chosen during the survey of HTLV- 1 carriers among American Indians. Genomic DNA from seven other American Indians (one Apache Indian, three Mayas, and three Brazilians) was provided by A. C. Wilson (University of California, Berkeley).

Study

Hutz, M. H., Callegari-Jacques, S. M. and Almeida, S. E. M. 2002. Low Levels of STRP Variability Are Not Universal in American Indians. Human Biology, 74(6): 791-806.

Study Summary

Data related to 15 short tandem repeat polymorphisms (STRPs) are reported for five Brazilian Indian populations, and a set of them compared with results previously reported for Asian, neo-Brazilian, North American, Iberian, and African populations. The low variability observed for these markers among the Surui Indians is confirmed, but the other populations show variability levels that are similar to those found elsewhere. Previous suggestions of population bottlenecks in the prehistorical colonization of the New World were not confirmed. On the other hand, STRPs again showed to be good markers for the establishment of population relationships.

Genetic Materials and Lines

Ricardo V. Santos and Carlos Coimbra Jr. provided the four American Indian samples. Fundação Nacional do Índio (FUNAI) gave permission to study the Indians. The Indian leaders and the subjects of the investigation were adequately informed about the aims of the study and gave their approval.

The Brazilian Indian samples, collected between 1988 and 1990, consisted of 146 individuals from the Gavião (29), Suruí (24), Zoró (30),Wai Wai (29), and Xavante (34) tribes. The Suruí sample was collected independently from that existing as lymphoblastoid cell lines; some individuals, however, could have been represented in both samples. The Gavião, Suruí, and Zoró populations are closely related, both culturally and geographically. They live in the southwestern region of the Brazilian Amazonia. The Wai-Wai Indians inhabit the northern Amazon, and the Xavante live in Central Brazil.

Study
Ingman, M. and Gyllensten, U. 2001. Analysis of the Complete Human mtDNA Genome: Methodology and Inferences for Human Evolution. The Journal of Heredity, 92(6): 454-461.

Study Summary
The analysis of mitochondrial DNA (mtDNA) sequences has been a potent tool in the researchers understanding of human evolution. However, almost all studies of human evolution based on mtDNA sequencing have focused on the control region, which constitutes less than 7% of the mitochondrial genome. The rapid development of technology for automated DNA sequencing has made it possible to study the complete mtDNA genomes in large numbers of individuals, opening the field of mitochondrial population genomics. Here the researchers describe a suitable methodology for determining the complete human mitochondrial sequence and the global mtDNA diversity in humans. Also, the researchers discuss the implications of the results with respect to the different hypotheses for the evolution of modern humans.

Genetic Materials and Lines
Not Available.

Study
Jin, L., Underwood, P. A., Doctor, V., Davis, R. W., Shen, P., Cavalli-Sforza, L. L. and Oefner, P. J. 1999. Distribution of Haplotypes from a Chromosome 21 Region Distinguishes Multiple Prehistoric Human Migrations. Proceedings of the National Academy of Science, 96: 3796-3800.

Study Summary
Despite mounting genetic evidence implicating a recent origin of modern humans, the elucidation of early migratory gene-flow episodes remains incomplete. Geographic distribution of haplotypes may show traces of ancestral migrations. However, such evolutionary signatures

can be erased easily by recombination and mutational perturbations. A 565-bp chromosome 21 region near the MX1 gene, which contains nine sites frequently polymorphic in human populations, has been found. It is unaffected by recombination and recurrent mutation and thus reflects only migratory history, genetic drift, and possibly selection. Geographic distribution of contemporary haplotypes implies distinctive prehistoric human migrations: one to Oceania, one to Asia and subsequently to America, and a third one predominantly to Europe. The findings with chromosome 21 are confirmed by independent evidence from a Y chromosome phylogeny. Loci of this type will help to decipher the evolutionary history of modern humans.

Genetic Materials and Lines

M.E. Ibrahim, T. Jenkins, J. Bertranpetit, P. Moral, T. Wagner, J. Kidd, M. Hsu, J. Chu, S.Q. Mehdi, P. Srinivasan, and P. Parham provided blood samples for use in this study.

Study

Kaestle, F. A. and Smith, D. G. 2001. Ancient mitochondrial DNA evidence for prehistoric population movement: The Numic expansion. American Journal of Physical Anthropology, 115(1): 1-12.

Study Summary

The mitochondrial DNA of modern American Indians has been shown to fall into one of at least five haplogroups (A, B, C, D, or X) whose frequencies differ among tribal groups. The frequencies of these five haplogroups in a collection of ancient individuals from Western Nevada dating to between approximately 350-9,200 years BP were determined. These data were used to test the hypothesis, supported by archaeological and linguistic data, that the current inhabitants of the Great Basin, the Numic speakers, are recent immigrants into the area who replaced the previous non-Numic inhabitants. The frequency distributions of haplogroups in the ancient and modern American Indians differed significantly, suggesting that there is a genetic discontinuity between the ancient inhabitants and the modern Numic speakers, providing further support for the Recent Numic Expansion hypothesis. The distribution of mitochondrial haplogroups of the ancient inhabitants of the Great Basin is most similar to those of some of the modern American Indian inhabitants of California.

Genetic Materials and Lines

The researchers thank the American Indians who donated blood and hair samples, and claim they granted permission for the study. They also thank Amy Dansie for her efforts in providing the ancient samples from Nevada.

Data for the modern American Indian groups studied were taken from the literature (Schurr et al., 1990; Torroni et al., 1992; Merriwether et al., 1995; Lorenz and Smith, 1996; Kaestle, 1997; Smith et al., 1999). Bone samples from 21 prehistoric individuals from the Pyramid Lake region and 27 prehistoric individuals from the Stillwater Marsh region of Western Nevada were obtained from the Nevada State Museum.

Study

Karafet, T., Zegura, S. L., Vuturo-Brady, J., Posukh, O., Osipova, L., Wiebe, V., Romero, F., Long, J., Harihara, S., Jin, F., Dashnyam, B., Gerelsaikhan, T., Keiichi, O. and Hammer, M. 1997. Y Chromosome Markers and Trans-Bering Strait Dispersals. American Journal of Physical Anthropology, 102: 301-314.

Study Summary

Five polymorphisms involving two paternally inherited loci were surveyed in 38 world populations (n=1,631) to investigate the origins of American Indians. One of the six Y chromosome combination haplotypes (1T) was found at relatively high frequencies (17.8-75.0%) in nine American Indian populations (n=206) representing the three major linguistic divisions in the New World. Overall, these data do not support the Greenberg et al. (1986) tripartite model for the early peopling of the Americas. The 1T haplotype was also discovered at a low frequency in Siberian Eskimos (3/22), Chukchi (1/6), and Evens (1/65) but was absent from 17 other Asian populations (n=987). The perplexing presence of the 1T haplotype in northeastern Siberia may be due to back-migration from the New World to Asia.

Genetic Materials and Lines

The researchers thank Chris Tyler-Smith, Mark Stoneking, William Klitz, Teri Markow, John Mitchell, Andrea Novelletto, Yehia Gad, Ling Ling Hsieh, A.V.S. Hill, Gudrun Rappold, Satoshi Horai, Gerald Shields, K. Tajima, S. Sonoda, V. Zaninovic, and Trefor Jenkins for donating DNA samples.

They analyzed a total of 1,631 males from 38 populations. Additionally, buccal cell DNA was prepared at the University of Arizona (where the sampling protocol was approved by the Human Subjects Committee) from 10 Tibetan monks, 26 southern Chinese students, and 42 Navajos associated with Navajo Community College were used. Other DNA samples were collected from 195 Mongolians, 41 Manchurian Evenks, 23 Chinese Oroqen, 30 Cheyenne, 24 Pima, 13 New Mexico Navajos, and 12 Tanana by various coauthors and their associates. DNA samples were provided by other investigators as follows: 21 British, 39 Italians, and 30 Egyptians by A. Novelletto; 7 Egyptians by Y. Gad; 42 Greeks and 11 British by J. Mitchell; 32 Germans by G. Rappold; 116 Japanese by S. Horai; 10 Havasupai and 15 Cheyenne by T. Markow; 15 Zapotecs by W. Klitz; 23 Wayus from Colombia by K. Tajima, S. Sonoda, and V. Zaninovic; 12 Evens, 12 Koryaks, and 4 Inupiaq Eskimos by G. Shields; 31 Australian Aboriginal People, and 36 Papua New Guineans by M. Stoneking; 48 Gambians by A.V.S. Hill; and 48 Bantu speakers from South Africa by T. Jenkins.

Study

Karafet, T. M., Osipova, L. P. and Gubina, M. A. 2002. High Levels of Y-Chromosome Differentiation among Native Siberian Populations and the Genetic Signature of a Boreal Hunter-Gatherer Way of Life. Human Biology, 74(6): 761-89.

Study Summary

The researchers examined genetic variation on the nonrecombining portion of the Y chromosome (NRY) to investigate the paternal population structure of indigenous Siberian groups and to reconstruct the historical events leading to the peopling of Siberia. A set of 62 biallelic markers on the NRY were genotyped in 1432 males representing 18 Siberian populations, as well as nine populations from Central and East Asia and one from European Russia. A subset of these markers defines the 18 major NRY haplogroups (A-R) recently described by the Y Chromosome Consortium (YCC 2002). While only four of these 18 major NRY haplogroups accounted for ~95 percent of Siberian Y-chromosome variation, native Siberian populations differed greatly in their haplogroup composition and exhibited the highest OST value for any region of the world. When the researchers divided the researchers Siberian sample into four geographic regions versus five major linguistic groupings, analyses of molecular variance (AMOVA) indicated higher OST and OCT values for linguistic groups than for geographic groups. Mantel tests also supported the existence of NRY genetic patterns that were correlated with language, indicating that language affiliation might be a better predictor of the genetic affinity among Siberians than their present geographic position. The combined results, including those from a nested cladistic analysis, underscored the important role of directed dispersals, range expansions, and long-distance colonizations bound by common ethnic and linguistic affiliation in shaping the genetic landscape of Siberia. The Siberian pattern of reduced haplogroup diversity within populations combined with high levels of differentiation among populations may be a general feature characteristic of indigenous groups that have small effective population sizes and that have been isolated for long periods of time.

Genetic Materials and Lines

The researchers analyzed 62 binary polymorphisms on the Y chromosomes of 902 males from 18 Siberian populations. Ten additional populations thought to have had contacts with Siberian groups were also included in the researchers analyses. Thus, the researchers total sample comprises 1432 Y chromosomes from 28 populations. These 28 populations were divided into six regional groupings based on arbitrary geographic criteria as follows: (1) Northwest Siberia (10 groups); (2) Central-South Siberia (6 groups); (3) Northeast Siberia (4 groups); (4) Central Asia (5 groups); (5) East Asia (2 groups); and (6) European Russia (1 group). New samples from the Khants, Komi, Nganasans, Dolgans, and Entsi were collected by T. Karafet and L. P. Osipova of the Institute of Cytology and Genetics in Novosibirsk with informed consent during 1999-2000 in the Yamal-Nenets and Taymyr Autonomous Districts. Additional genomic DNAs from the Kets, Forest Nentsi, and Yakut-Sakha populations were collected by the laboratory of L. P. Osipova in Russia. All samples were collected in traditional settlements. Demographic and pedigree data were obtained along with blood samples. From demographic and genealogical information the researchers were able to identify paternally unrelated males (for at least three to six generations). All sampling protocols were approved by the Human Subjects Committee at the University of Arizona.

Study

Karafet, T. M., Zegura, S. L., Posukh, O., Osipova, L., Bergen, A., Long, J., Goldman, D., Klitz, W., Harihara, S., de Knijff, P., Wiebe, V., Griffiths, R. C., Templeton, A. R. and Hammer, M. F. 1999. Ancestral Asian Source(s) of New World Y-Chromosome Founder Haplotypes. American Journal of Human Genetics, 64: 817-831.

Study Summary

Haplotypes constructed from Y-chromosome markers were used to trace the origins of American Indians. The researchers sample consisted of 2,198 males from 60 global populations, including 19 American Indian and 15 indigenous North Asian groups. A set of 12 biallelic polymorphisms gave rise to 14 unique Y-chromosome haplotypes that were unevenly distributed among the populations. Combining multiallelic variation at two Y-linked microsatellites (DYS19 and DXYS156Y) with the unique haplotypes results in a total of 95 combination haplotypes. Contra previous findings based on Y-chromosome data, their new results suggest the possibility of more than one American Indian parental founder haplotype. They postulate that, of the nine unique haplotypes found in American Indians, haplotypes 1C and 1F are the best candidates for major New World founder haplotypes, whereas haplotypes 1B, 1I, and 1U may either be founder haplotypes and/or have arrived in the New World via recent admixture. Two of the other four haplotypes (YAP+ haplotypes 4 and 5) are probably present because of post-Columbian admixture, whereas haplotype 1G may have originated in the New World, and the Old World source of the final New World haplotype (1D) remains unresolved. The contrasting distribution patterns of the two major candidate founder haplotypes in Asia and the New World, as well as the results of a nested cladistic analysis, suggest the possibility of more than one paternal migration from the general region of Lake Baikal to the Americas.

Genetic Materials and Lines

The researchers analyzed a total of 2,198 males from 60 worldwide populations. The DNA samples included subsets of the samples examined by Hammer et al. (1997, 1998) and Karafet et al. (1997), although the exact number of subjects for each population occasionally varies among these studies. In addition, they included the following new samples: 62 Inuit Eskimos, 12 Mixe, 29 Mixtecs, 22 Kazakhs, 30 Evenks, and 18 Melanesians, which were collected by the coauthors, whereas 17 Ngobe, 12 Kuna, 10 Embera, and 15 Wounan from Panama were provided by E. Bermingham and C. Kolman. All sampling protocols were approved by the Human Subjects Committee at the University of Arizona.

Study

Keyeux, G., Rodas, C., Gelvez, N. and Carter, D. 2002. Possible migration routes into South America deduced from mitochondrial DNA studies in Colombian Amerindian populations. Human Biology, 74(2): 211-33.

Study Summary

Mitochondrial DNA haplotype studies have been useful in unraveling the origins of American Indians. Such studies are based on restriction site and intergenic deletion/insertion polymorphisms, which define four main haplotype groups common to Asian and American populations. Several studies have characterized these lineages in North, Central, and South American Amerindian, as well as Na Dene and Aleutian populations. Siberian, Central Asian, and Southeast Asian populations have also been analyzed, in the hope of fully depicting the route(s) of migration between Asia and America. Colombia, a key route of migration between North and South America, has until now not been studied. To resolve the current lack of information about Colombian Amerindian populations, the researchers have investigated the presence of the founder haplogroups in 25 different ethnic groups from all over the country. The present research is part of an interdisciplinary program, Expedicion Humana, fostered by the Universidad Javeriana and Dr. J. E. Bernal V. The results show the presence of the four founder A-D Amerindian lineages, with varied distributions in the different populations, as well as the presence of other haplotypes in frequencies ranging from 3% to 26%. These include some unique or private polymorphisms, and also indicate the probable presence of other Asian and a few non-Amerindian lineages. A spatial structure is apparent for haplogroups A and D, and to a lesser extent for haplogroup C. While haplogroup A and D frequencies in Colombian populations from the northwestern side of the Andes resemble those seen in Central American Amerindians more than those seen in South American populations, their frequencies on the southeastern side more closely resemble the bulk of South American frequencies so far reported, raising the question as to whether they reflect more than one migration route into South America. High frequencies of the B lineage are also characteristic of some populations. Their observations may be explained by historical events during the pre-Columbian dispersion of the first settlers and, later, by disruptions caused by the European colonization.

Genetic Materials and Lines

Drs. Mark Stoneking and Thomas White supported this study by providing the primers and donating the Taq polymerase. Dr. Agnes Rötig provided the Leber primers. The researchers thank the people of Colombia who voluntarily contributed to these insights into human history.

Blood samples from 681 unrelated individuals belonging to the 25 different Amerindian groups outlined above were collected under the auspices of "Expedición Humana," an interdisciplinary research program fostered by the Javeriana University. All participants, as well as the communities' hierarchical leaders, gave their free consent.

Study

Lalueza-Fox, C., Gilbert, M. T. P., Martinez-Fuentes, A. J., Calafell, F. and Bertranpetit, J. 2003. Mitochondrial DNA from pre-Columbian Ciboneys from Cuba and the prehistoric colonization of the Caribbean. American Journal of Physical Anthropology, 121(2): 97-108.

Study Summary

To assess the genetic affinities of extinct Ciboneys (also called Guanajuatabeys) from Cuba, 47 pre-Columbian skeletal samples belonging to this group were analyzed using ancient DNA techniques. At the time of European contact, the center and east of Cuba were occupied by agriculturalist Taino groups, while the west was mainly inhabited by Ciboneys, hunter-gatherers who have traditionally been considered a relic population descending from the initial colonization of the Caribbean. The mtDNA hypervariable region I (HVR-I) and haplogroup-specific markers were amplified and sequenced in 15 specimens using overlapping fragments; amplification from second extractions from the same sample, independent replication in different laboratories, and cloning of some PCR products support the authenticity of the sequences. Three of the five major mtDNA Amerindian lineages (A, C, and D) are present in the sample analyzed, in frequencies of 0.07, 0.60, and 0.33, respectively. Different phylogenetic analyses seem to suggest that the Caribbean most likely was populated from South

America, although the data are still inconclusive, and Central American influences cannot be discarded. The researchers hypothesis is that the colonization of the Caribbean mainly took place in successive migration movements that emanated from the same area in South America, around the Lower Orinoco Valley: the first wave consisted of hunter-gatherer groups (ancestors of the Ciboneys), a subsequent wave of agriculturalists (ancestors of the Tainos), and a latter one of nomadic Carib warriors. However, further genetic studies are needed to confirm this scenario.

Genetic Materials and Lines

Forty-seven samples from Ciboney culture sites were analyzed. The samples belong to three different sites: Perico I cave (N=37), 25 km west of Bahia Honda, in Pinar del Rio (Cuba), Mogote La Cueva (N=3), and Canimar (N=7). Perico cave, one of the best studied preagriculturalist sites, has a radiocarbon dating of 1,990+/- 50 BP (unpublished data). The site was excavated in 1970 and 1997, and yielded the remains of at least 162 individuals (Travieso Ruiz et al., 1999). Mogote La Cueva is a site in Pinar del Rio, and has been radiocarbon-dated at 1,620 BP (SI-424) (Tabio and Rey, 1966); Canimar is a rock shelter over the Canimar River (Matanzas), dated to 4,700+/-70 BP (UBAR-171) (unpublished data). The specimens were chosen from a wider sample due to their good external preservation, taking in consideration a "fresh" aspect, lack of mineralization, absence of cracks or bone erosion, and completeness of the specimens. The skeletal material is held at the Museo Antropologico Montane (Facultad de Biologia, Universidad de La Habana, Havana, Cuba).

Study

Lell, J.T., Brown, M.D., Schurr, T.G., Sukernik, R.I., Starikovskaya, Y.B., Torrnoi, A., et al. 1997. Y Chromosome Polymorphisms in Native American and Siberian Populations: Identification of Native American Y Chromosome Haplotypes. Human Genetics, 100(5-6): 8.

Study Summary
Not Available.

Genetic Materials and Lines
Not Available.

Study
Lell, J. T., Sukernik, R. I., Starikovskaya, Y. B., Su, B., Jin, L., Schurr, T. G., Underhill, P. A. and Wallace, D. C. 2002. The dual origin and Siberian affinities of American Indian Y chromosomes. American Journal of Human Genetics, 70(1): 192-206.

Study Summary
The Y chromosomes of 549 individuals from Siberia and the Americas were analyzed for 12 biallelic markers, which defined 15 haplogroups. The addition of four microsatellite markers increased the number of haplotypes to 111. The major American Indian founding lineage, haplogroup M3, accounted for 66% of male Y chromosomes and was defined by the biallelic markers M89, M9, M45, and M3. The founder haplotype also harbored the microsatellite alleles DYS19 (10 repeats), DYS388 (11 repeats), DYS390 (11 repeats), and DYS391 (10 repeats). In Siberia, the M3 haplogroup was confined to the Chukotka peninsula, adjacent to Alaska. The second major group of American Indian Y chromosomes, haplogroup M45, accounted for about one-quarter of male lineages. M45 was subdivided by the biallelic marker M173 and by the four microsatellite loci alleles into two major subdivisions: M45a, which is found throughout the Americas, and M45b, which incorporates the M173 variant and is concentrated in North and Central America. In Siberia, M45a haplotypes, including the direct ancestor of haplogroup M3, are concentrated in Middle Siberia, whereas M45b haplotypes are found in the Lower Amur River and Sea of Okhotsk regions of eastern Siberia. Among the remaining 5 percent of American Indian Y chromosomes is haplogroup RPS4Y-T, found in North America. In Siberia, this haplogroup, along with haplogroup M45b, is concentrated in the Lower Amur River/Sea

of Okhotsk region. These data suggest that American Indian male lineages were derived from two major Siberian migrations. The first migration originated in southern Middle Siberia with the founding haplotype M45a (10-11-11-10). In Beringia, this gave rise to the predominant American Indian lineage, M3 (10-11-11-10), which crossed into the New World. A later migration came from the Lower Amur/Sea of Okhotsk region, bringing haplogroup RPS4Y-T and subhaplogroup M45b, with its associated M173 variant. This migration event contributed to the modern genetic pool of the Na-Dene and Amerinds of North and Central America.

Genetic Materials and Lines

All of the samples analyzed in the study were originally collected as whole blood. Most of the population samples presented are subsets of previously studied population sets: southern Mexico American Indians (Torroni et al. 1994); Navajos (Torroni et al. 1992); Seminoles (Huoponen et al. 1997); Siberian Eskimos and Chukchi (Starikovskaya et al. 1998); Koryaks and Itelmens (Schurr et al. 1999); Yenisey Evenks, Udegeys, and Nivkhs (Torroni et al. 1993b); and Kets (Sukernik et al. 1996).

The samples representing southern Middle Siberia include the Turkic-speaking Tofalars and Tuvans, as well as the Mongolic-speaking Buryats. The samples from the Russian Far East were the Tungusic-speaking Ulchi/Nania, Negidals, and Okhotsk Evenks. These population aggregates of the Altaic linguistic family are located on opposite sides of the southern Siberian belt. The remnants of nonadmixed Tofalars, a small tribe of hunters and reindeer breeders occupying the Taiga area on the northern slopes of the eastern Sayan mountain range (Levin and Potapov 1964), were sampled in the villages of Alygdzher, Nerkha, and Upper Gutara, all of which are in the Nizhneudinsk administrative district of the Irkutsk Region. The Tuvan samples were collected in Kizil, the capital of the Tuva Republic, mainly from students of Tuva University. Buryat samples were collected in Kushun village, Nizhneudinsk District, Irkutsk Region. These samples represent the Buryats of the Sayan-Baikal upland. Blood samples from Ulchi and Nanai individuals were collected in the villages of Old and New Bulava in the Ulchi District of the Khabarovsk Region. Samples from a geographically isolated group of Evenks were collected in several small settlements on the mainland Okhotsk Sea shore in the Tugur-Chumikan District of the Khabarovsk Region. Additionally, remnants of the Negidals, who were swamped by the expanding Evenks, were sampled in the Polina Osipenko District (Upriver Negidals) and the Nikolayevski District (Downriver Negidals) of the Khabarovsk Region. Many of the Central American and all of the South American samples were collected and supplied by J. V. Neel.

Study

Lorenz, J. and Smith, D. G. 1996. Distribution of Four Founding mtDNA Haplogroups Among Native North Americans. American Journal of Physical Anthropology, 101: 307-323.

Study Summary

The mtDNA of most American Indians has been shown to cluster into four lineages, or haplogroups. This study provides data on the haplogroup affiliation of nearly 500 Native North Americans including members of many tribal groups not previously studies. Phenetic cluster analysis shows a fundamental difference among 1) Eskimos and northern Na-Dene groups, which are almost exclusively mtDNA haplogroups A; 2) tribes of the Southwest and adjacent regions, predominantly Hokan and Uto-Aztecan speakers, which lack haplogroup A but exhibit high frequencies of haplogroup B; 3) tribes of the Southwest and Mexico lacking only haplogroup D; and 4) a geographically heterogonous group of tribes which exhibit varying frequencies of all four haplogroups. There is some correspondence between language group affiliations and the frequencies of the mtDNA haplogroups in certain tribes, while geographic proximity appears responsible for the genetic similarity among other tribes. Other instances of similarity among tribes suggest hypotheses for testing with more detailed studies.

Genetic Materials and Lines
Indian Health Services Facilities provided much of the data. Dr. John Johnson (Santa Barbara Museum of Natural History) and Mr. David L. Schmidt (University College of Cape Brown, Nova Scotia) provided hair samples.

The haplogroup affiliation of 497 individuals from more than 40 ethnic groups across North America was determined. The source and description of these samples have been presented elsewhere (Lorenz and Smith, 1994).

Study
Lorenz, J. G. and Smith, D. G. 1994. Distribution of the 9-bp mitochondrial DNA region V deletion among North American Indians. Human Biology; An International Record Of Research, 66(5): 777-788.

Study Summary
The deletion of a 9-bp segment from the intergenic region between the mtDNA cytochrome oxidase II gene and the lysine tRNA gene has been documented mainly in individuals of East Asian ancestry and in individuals from East Asian-derived populations (e.g., Polynesia). Among American Indians the deletion is absent among Eskimos and northern Na-Dene populations and present among most Amerind populations [sensu Greenberg (1987); i.e., all American Indians except Eskimo-Aleut and Na-Dene] that have been studied. To better characterize the frequency and distribution of the 9-bp deletion in North America, the researchers surveyed more than 400 individuals from 59 tribes representing a variety of linguistic groups. The absence of the deletion among Eskimo and northern Na-Dene populations is confirmed. Among Amerind groups the deletion is present in all groups represented by more than six individuals. The geographic distribution of the frequencies of the deletion appears to be clinal in North America. The deletion is absent in the Artic and Subartic and reaches its highest frequency in the Southwest. This distribution is consistent with the hypothesis that the ancestors of the Amerinds and Na-Dene arrived in the New World by means of separate migrations. The presence of the 9-bp deletion in high

frequencies in all the major linguistic groups in the Southwest suggests that migration among tribes was common.

Genetic Materials and Lines

The samples used for this study were randomly selected from more than 3500 samples that were collected for clinical studies unrelated to this study and are predominantly represented by serum. Genomic DNA was extracted from whole blood samples drawn from individuals representing the Kumiai, Cocopa, Paipai, Kiliwa, and Cochimi groups and from hair samples obtained from individuals representing most of the California groups. When the ethnic background is reported, only individuals that allege to be "fullblood" were included. Individuals represented by the hair samples were determined to be matrilineally "fullblood" by genealogical records. For the California and Baja Yuman samples only maternally unrelated individuals were included in this study.

The following individuals provided samples: Peter Bennett, National Institutes of Health, Phoenix, Arizona; Lyle Best, PHS Indian Hospital, Belcourt, North Dakota; Michelle Boudin, Lac Courte Oreilles Community Health Center, Hayward, Wisconsin; Jerome DeWolfe, Fort Totten Health Center, Fort Totten, North Dakota; Robert Ferrell, Graduate School of Public Health, University of Pittsburgh, Pittsburgh, Pennsylvania; L. Leigh Field, Division of Medical Genetics, University of Calgary, Calgary, Canada; Douglas Forman, Chickasaw Nation Health Clinic, Ardmore, Oklahoma; Henry Gershowitz, Department of Human Genetics, University of Michigan, Ann Arbor; Vincent Henderson, PHS Indian Hospital, Concho, Oklahoma; John Johnson, Santa Barbara Museum of Natural History, Santa Barbara, California; D. Kasprisin, American Red Cross, Tulsa, Oklahoma; Gert Lambert, Ne-ia Shing IHS Clinic, Onamia, Minnesota; Ruben Lisker, National Institute of Nutrition, Mexico City, Mexico; Gerald Ross, Jemez Indian Health Center, Jemez Pueblo, New Mexico; Patricia Samuelson, Sacramento Urban Indian Health Project, Sacramento, California; David L. Schmidt, University of California, Davis; Arthur Steinberg, Cleveland Clinic Foundation, Cleveland, Ohio; Beverly Stone, Cherokee Nation Indian Health Clinic, Stillwell, Oklahoma; Hector Velasquez,

University of Baja, California, Medical School, Mexicali, Mexico.

Study

Lorenz, J. G. and Smith, D. G. 1997. Distribution of sequence variation in the mtDNA control region of native North Americans. Human Biology, 69(6): 749-776.

Study Summary

The distributions of mtDNA diversity within and/or among North American haplogroups, language groups, and tribes were used to characterize the process of tribalization that followed the colonization of the New World. Approximately 400 bp from the mtDNA control region of 1 Na-Dene and 33 Amerind individuals representing a wide variety of languages and geographic origins were sequenced, With the inclusion of data from previous studies, 225 native North American (284 bp) sequences representing 85 distinct mtDNA lineages were analyzed. Mean pairwise sequence differences between (and within) tribes and language groups were primarily due to differences in the distribution of three of the four major haplogroups that evolved before settlement of the New World. Pairwise sequence differences within each of these three haplogroups were more similar than previous studies based on restriction enzyme analysis have indicated. The mean of pairwise sequence differences between Amerind members of haplogroup A, the most common of the four haplogroups in North America, was only slightly higher than that for the Eskimo, providing no evidence of separate ancestry, but was about two-thirds higher than that for the Na-Dene. However, analysis of pairwise sequence divergence between only tribal-specific lineages, unweighted for sample size, suggests that random evolutionary processes have reduced sequence diversity within the Na-Dene and that members of all three language groups possess approximately equally diverse mtDNA lineages. Comparisons of diversity within and between specific ethnic groups with the largest sample size were also consistent with this outcome. These data are not consistent with the hypothesis that the New World was settled by more than a single migration. Because Lineages tended not to cluster by tribe and because lineage sharing among linguistically unrelated groups was restricted to geographically proximate groups, the tribalization process probably did not occur soon after settlement of the New World, and/or considerable admixture has occurred among daughter populations.

Genetic Materials and Lines

All sequence analysis was performed on DNA that had been extracted for earlier studies (Lorenz and Smith 1994, 1996) and assigned to haplogroups A, B, or C. Members of haplogroup D, whose distribution in North America is more limited, were not included in the present study because they lack unique markers in the control region and therefore cannot be identified by their published control region sequence. The ethnic affiliations of the 34 individuals whose mtDNAs were sequenced in this study are as follows: 5 Nahuatl, 5 Chumash, 2 Bella Coola, 1 Choctaw, 2 Washo, 2 Ojibwa (Chippewa), 1 Dogrib, 1 Hopi, 1 Salinan, 4 California Uto-Aztecans (represented by 1 individual each from the Cahuilla, Fernandeno, Vanyume, and Serreno tribes), and 10 Yuman speakers from Baja, Mexico (represented by 5 Kumiai, 2 Paipai, and 1 individual each from the Kiliwa, Cucapa, and Cochimi tribes).

Study

Malhi, R.S., Eshleman, J.A., Greenberg, J.A., Weiss, D.A., Schultz Shook, B.A., Kaestle, F.A., et al. 2002. The Structure of Diversity within New World Mitochondrial DNA Haplogroups: Implications for the Prehistory of North America. American Journal of Human Genetics, 70(4): 905-919.

Study Summary

The mitochondrial DNA haplogroups and hypervariable segment I (HVSI) sequences of 1,612 and 395 Native North Americans, respectively, were analyzed to identify major prehistoric population events in North America. Gene maps and spatial autocorrelation analyses suggest that populations with high frequencies of haplogroups A, B, and X experienced prehistoric population expansions in the North,

Southwest, and Great Lakes region, respectively. Haplotype networks showing high levels of reticulation and high frequencies of nodal haplotypes support these results. The haplotype networks suggest the existence of additional founding lineages within haplogroups B and C; however, because of the hypervariability exhibited by the HVSI data set, similar haplotypes exhibited in Asia and America could be due to convergence rather than common ancestry. The hypervariability and reticulation preclude the use of estimates of genetic diversity within haplogroups to argue for the number of migrations to the Americas.

Genetic Materials and Lines
Not Available.

Study
Malhi, R. S., Breece, K. E., Shook, B. A., Kaestle, F. A., Chatters, J. C., Hackenberger, S. and Smith, D. G. 2004. Patterns of mtDNA Diversity in Northwestern North America. Human Biology, 76(1): 33-54.

Study Summary
The mitochondrial DNA (mtDNA) haplogroups of 54 fullblooded modern and 64 ancient American Indians from northwestern North America were determined. The control regions of 10 modern and 30 ancient individuals were sequenced and compared. Within the Northwest, the frequency distribution for haplogroup A is geographically structured, with haplogroup A decreasing with distance from the Pacific Coast. The haplogroup A distribution suggests that a prehistoric population intrusion from the subarctic and coastal region occurred on the Columbia Plateau in prehistoric times. Overall, the mtDNA pattern in the Northwest suggests significant amounts of gene flow among Northwest Coast, Columbia Plateau, and Great Basin populations.

Genetic Materials and Lines
The sources for samples from the Bella Coola (a coastal Salish group) and the Nuu-Chah-Nulth are described

by Smith, Lorenz et al. (2000). The Vantage group is represented by tooth samples from the Middle Columbia River and is radiocarbon dated to 500–1500 yr b.p. (S. Hackenberger, unpublished data, 1999). The protohistoric tooth samples from San Poil, Nespelem, Wenatchee, and Douglas County are approximately 200 years old and were found in regions occupied by speakers of Salish languages in prehistoric times. The protohistoric tooth samples from the Snake River, John Day River, and Palouse River are approximately 200 years old and were collected in regions occupied by Sahaptian speakers in prehistoric times. Five protohistoric tooth samples known only to be from the general Columbia Plateau region were also analyzed. The samples were provided by Central Washington University, Ellensburg. Tooth samples from the protohistoric Memaloose Island cemetery, near the Dalles, were provided by Yale University and were also analyzed as part of this study; they are estimated to date to 200 yr b.p. Individuals from this population were probably related to the Chinookan-speaking Wishram, who lived nearby on the north side of the Columbia River. The haplogroups of 54 full-blooded modern and 64 ancient American Indians were determined by restriction fragment length polymorphism (RFLP) analysis. A subset of 10 modern and 30 ancient samples was sequenced (from np 16,055 to np 16,548 for modern samples and from np 16,055 to np 16,356 for ancient samples).

Study

Malhi, R. S., Schultz, B. A. and Smith, D. G. 2001. Distribution of mitochondrial DNA lineages among American Indian tribes of northeastern North America. Human Biology, 73(1): 17-55.

Study Summary

The mtDNA haplogroups of 185 individuals from American Indian tribes in Northeast North America were determined. A subset of these individuals was analyzed by sequencing hypervariable segments I and II of the control region. The haplogroup frequency distributions of populations in the Northeast exhibit regional continuity that predates

European contact. A large amount of gene flow has occurred between Siouan- and Algonquian-speaking groups, probably due to an Algonquian intrusion into the Northeast. The data also support both the Macro-Siouan hypothesis and a relatively recent intrusion of Northern Iroquoians into the Northeast. These conclusions are consistent with archaeological and linguistic evidence.

Genetic Materials and Lines

The Northern Ontario Ojibwa located north of Lake Superior in Ontario, whose mtDNA haplogroups were analyzed by Scozzari et al. (1997), were originally studied by Szathmary et al. (1978) and Torroni et al. (1993). The Manitoulin Island (Central) Ojibwa, originally described by Szathmary et al. (1978), were also studied by Scozzari et al. (1997). The Turtle Mountain Chippewa samples were collected by Lorenz and Smith (1996) from the Public Health Service Hospital in Belcourt, North Dakota.

The Wisconsin Chippewa samples were collected by Lorenz and Smith (1996) from an Indian Health Services clinic in Hayward, Wisconsin, serving a population that speaks the Southwestern Ojibwa dialect. The Micmac samples were collected for Lorenz and Smith (1996) by Dr. David L. Schmidt from residents of Eskasoni, Nova Scotia.

The Cheyenne/Arapaho samples were collected by Lorenz and Smith (1996) from the Indian Health Center in Concho, Oklahoma. Since these Cheyenne and Arapaho are both closely related and (recently) highly admixed with each other, the researchers have treated them as a single sample.

Samples of Sisseton/Wapheton Sioux, a Central Siouan group, were collected by Lorenz and Smith (1996) from the Fort Totten Health Center in Fort Totten, North Dakota. The Norris Farms (Oneota) samples were extracted from skeletal remains collected from a pre-Columbian (approx. 700 ybp) cemetery site located in west-central Illinois and analyzed by Stone and Stoneking (1993; 1998). Based on the geographic location of this site, its inhabitants are assumed to be ancestors of modern Siouan groups.

The Mohawk samples were collected and analyzed by Merriwether and Ferrell (1996). The Stillwell Cherokee samples were collected by Lorenz and Smith (1996) from the Cherokee Nation Indian Health Clinic in Stillwell, Oklahoma. The Oklahoma Red Cross Cherokee samples were provided to Lorenz and Smith (1996) by the American Red Cross in Tulsa, Oklahoma, who collected them from Cherokee individuals at various clinics throughout the designated Cherokee district in Oklahoma. Finally, the Pawnee samples were collected by Lorenz and Smith (1996) from individuals that reside on the Pawnee reservation in Oklahoma.

Study
Malhi, R.S., and Smith, D.G. 2002. Brief Communication: Haplogroup X Confirmed in Prehistoric North America. American Journal of Physical Anthropology, 119(1): 84-86.

Study Summary
Haplogroup X represents approximately 3% of all modern Native North American mitochondrial lineages. Using RFLP and hypervariable segment I (HVSI) sequence analyses, the researchers identified a prehistoric individual radiocarbon dated to 1,340 +/- 40 years BP that is a member of haplogroup X, found near the Columbia River in Vantage, Washington. The presence of haplogroup X in prehistoric North America, along with recent findings of haplogroup X in southern Siberians, confirms the hypothesis that haplogroup X is a founding lineage.

Genetic Materials and Lines
Not Available.

Study
Martinez-Cruzado, J. C., Toro-Labrador, G., Ho-Fung, V., Estevez-Montero, M. A., Lobaina-Manzanet, A., Padovani-Claudio, D. A., Sanchez-Cruz, H., Ortiz-Bermudez, P. and Sanchez-Crespo, A. 2001. Mitochondrial DNA analysis reveals substantial American Indian ancestry in Puerto Rico. Human Biology, 73(4): 491-511.

Study Summary

To estimate the maternal contribution of American Indians to the human gene pool of Puerto Ricans — a population of mixed African, European, and Amerindian ancestry — the mtDNAs of two sample sets were screened for restriction fragment length polymorphisms (RFLPs) defining the four major American Indian haplogroups. The sample set collected from people who claimed to have a maternal ancestor with American Indian physiognomic traits had a statistically significant higher frequency of American Indian mtDNAs (69.6%) than did the unbiased sample set (52.6%). This higher frequency suggests that, despite the fact that the native Taino culture has been extinct for centuries, the Taino contribution to the current population is considerable and some of the Taino physiognomic traits are still present. American Indian haplogroup frequency analysis shows a highly structured distribution, suggesting that the contribution of American Indians foreign to Puerto Rico is minimal. Haplogroups A and C cover 56.0% and 35.6% of the American Indian mtDNAs, respectively. No haplogroup D mtDNAs were found. Most of the linguistic, biological, and cultural evidence suggests that the Ceramic culture of the Tainos originated in or close to the Yanomama territory in the Amazon. However, the absence of haplogroup A in the Yanomami suggests that the Yanomami are not the only Taino ancestors.

Genetic Materials and Lines

Rob Donnelly (New Jersey Medical School Molecular Resource Facility) provided the oligonucleotide primers used in this research.

Two different sample sets were obtained following informed consent. The first sample set was biased for American Indian ancestry. It included 56 Puerto Ricans living in communities known historically for their strong Indian component, or those not living in these communities but having mothers or maternal grandmothers with phenotypic traits identified by themselves as "Indianlike," such as straight, dark hair; pronounced cheekbones; almond-shaped, dark eyes; and bronzed skin color. Of these persons, 10 came from Sector el

Treinta, Indiera Alta, Maricao; 8 from Indiera Baja, Maricao; and 5 from Miraflores, Añasco. Another 33 persons had mothers or maternal grandmothers with Indian-like phenotypic traits or who had originated from central Puerto Rico, an area that is considered to have the largest degree of Indian ancestry. This last group was composed of people associated with the University of Puerto Rico at Mayagüez or living in western Puerto Rico. The second, unbiased sample set was obtained independently from the first. It was taken at random from 38 Puerto Ricans associated with the University of Puerto Rico at Mayagüez or living in Mayagüez. In discussing comparisons between them, the last group of the biased sample set and the unbiased sample set will be referred to as the American Indian–biased university group and the unbiased sample group, respectively.

Study

Martinez-Laso, J., Sartakova, M., Allende, L., Konenkov, V., Moscoso, J., Silvera-Redondo, C., Pacho, A., Trapaga, J., Gomez-Casado, E. and Arnaiz-Villena, A. 2001. HLA molecular markers in Tuvinians: a population with both Oriental and Caucasoid characteristics. Annals of Human Genetics, 65: 245-261.

Study Summary

HLA class I and class II alleles have been studied for the first time in the Turkish-speaking Tuvinian population, which lives in Russia, North of Mongolia and close to the Altai mountains. Comparisons have been done with about 11000 chromosomes from other worldwide populations, and extended haplotypes, genetic distances, neighbor joining dendrograms and correspondence analyses have been calculated. Tuvinians show an admixture of Mongoloid and Caucasoid characters; the latter probably coming from the ancient Kyrgyz background or, less feasibly, more recent Russian Caucasoid admixture. However, Siberian population traits are not found and thus Tuvinians are closer to Central Asian populations. Siberians are more related to Na-Dene and Eskimo American Indians; Amerindians (from nowadays Iberian-America) are not related to any other group, including Pacific Islanders, Siberians or other American Indians. The more than one wave model for the peopling of the Americas is supported.

Genetic Materials and Lines

One-hundred-and-ninety healthy unrelated Tuvinians, all permanent residents of Kyzyl in the Republic of Tuva and with at least two recorded generations of Tuvinian descent were studied for HLA genotyping and phylogenetical calculations. The samples were taken by V. I. Konenkov and M. Sartakova from Institute of Clinical Immunology, Novosibirsk, Russia. A total of 11144 chromosomes were studied, including populations from different ethnic backgrounds: Caucasoids, Orientals, Polynesians, Micronesians, Na-Dene, Eskimos, and Amerindians.

Study

Merriwether, A., Hell, W. W., Vahlne, A. and Ferrell, R. E. 1996. mtDNA Variation Indicates Mongolia May Have Been the Source for the Founding Population for the New World. American Journal of Human Genetics, 59: 204-212.

Study Summary

mtDNA RFLP variation was analyzed in 42 Mongolians from Ulan Bator. All four founding lineage types (A [4.76%], B [2.38%], C [11.9%], and D [19.04%]) identified by Torroni and colleagues were detected. Seven of the nine founding lineage types proposed by Bailliet and colleagues and Merriwether and Ferrell were detected (A2 [4.76%], B [2.38%], C1 [11.9%], D1 [7.14%], D2 [11.9%], X6 [16.7%], and X7 [9.5%]). Sixty-four percent of these 42 individuals had "Amerindian founding lineage" haplotypes. A survey of 24 restriction sites yielded 16 polymorphic sites and 21 different haplotypes. The presence of all four of the founding lineages identified by the Torroni group (and seven of Merriwether and Ferrell's nine founding lineages), combined with Mongolia's location with respect to the Bering Strait, indicates that Mongolia is a potential location for the origin of the founders of the New World. Since lineage B, which is widely distributed in the New World, is absent in Siberia, the researchers concluded that Mongolia or a geographic location common to both contemporary Mongolians and American aboriginals is the more likely origin of the founders of the New World.

Genetic Materials and Lines

DNA was extracted from the cell pellets of 45 Native Mongolians from the capital city of Ulan Bator in north-central Mongolia.

Study

Merriwether, D.A., Rothhammer, F., and Ferrell, R.E. 1994. Genetic Variation in the New World: Ancient Teeth, Bone, and Tissue as Sources of DNA. Experientia, 50(6): 592-601.

Study Summary

Examination of ancient and contemporary American Indian mtDNA variation via diagnostic restriction sites and the 9-bp Region V deletion suggests a single wave of migration into the New World. This is in contrast to data from Torroni et al., which suggested two waves of migration into the New World (the NaDene and Amerind). All four founding lineage types are present in populations in North, Central, and South America suggesting that all four lineages came over together and spread throughout the New World. Ancient American Indian DNA shows that all four lineages were present before European contact in North America, and at least two were present in South America. The presence of all four lineages in the NaDene and the Amerinds argues against separate migrations founding these two groups, although admixture between the groups is still a viable explanation for the presence of all four types in the NaDene.

Genetic Materials and Lines

Not Available.

Study

Merriwether, A., Rothhammer, F. and Ferrell, R. E. 1995. Distribution of the Four Founding Lineage Haplotypes in American Indians Suggests a Single Wave of Migration for the New World. American Journal of Physical Anthropology, 98: 411-430.

Study Summary

The distribution of the four founding lineage haplogroups in American Indians from North, Central, and South America shows a north to south increase in the frequency of lineage B and a north to south decrease in the frequency of lineage A. All four founding lineage haplogroups were detected in North, Central, and South America, and in Greenberg's et al's (1986) three major linguistic groups (Amerind, NaDene, and Eskaleut), with all four haplogroups often found within a single population. Lineage A was the most common lineage in North America, regardless of language group. This overall distribution is most parsimonious with a single wave of migration into the New World, which included multiple variants of all four founding lineage types. Torroni et al.'s (1993) report that lineage B has a more recent divergence time than the other three can best be explained by multiple variants of lineages A,C, and D, and fewer variants of B entering the New World. Alternatively, there could have been multiple waves of migration from a single parent population in Asia/Siberia, which repeatedly reintroduced the same lineages.

Genetic Materials and Lines

Dr. Emoke Szathmary provided Dogrib and (with Naomi Adelson) Mohawk samples. Dr. Kenneth Weiss provided the Mvskoke and Mayan samples, and Dr. William S. Laughlin for providing the St. Lawrence and Kodiak Island Eskimo and Aleut samples.

Mvskoke Indians (n=71) (Russell et al., 1994) from Muskogee, Oklahoma, were provided by Dr. Kenneth Weiss. Mohawk (n=208) from eastern-central Canada were provided by Dr. Emoke Szathmary and Dr. Naomi Adelson. Dogrib (n=169) from northwest Canada (Szathmary et al., 1983) were collected by Dr. Emoke Szathmary in 1980 and 1985. Alaskan

Eskimo samples were collected from Kodiak Island [Old Harbor (n=156) and Ouzinkie (n=56)] (Majumder et al., 1988), St. Lawrence Island [Savoonga (n=68) and Gambell (n=72)] (Ferrell et al., 1981), and Southwestern Alaska [Yupik Eskimos (n=177)] (Petersen et al., 1991). Alaskan Aleut samples were collected from Pribilof Island from the village of St. Paul (n=78) (Majumder et al., 1988).

Study
Merriwether, D. A. and Ferrell, R. E. 1996. The four founding lineage hypothesis for the New World: a critical reevaluation. Molecular Phylogenetics and Evolution, 5(1): 241-6.

Study Summary
It has been proposed that all American Indian mitochondrial DNA variation in the New World can be attributed to divergence from four "founding lineages" which entered the New World in three waves of migration from across the Bering Strait. Torroni et al. (1993a) believe that only one haplotype from each of these four founding lineages arrived in the New World via migration, and all the additional variation arose in the New World. Any other types were attributed to Caucasian admixture. G. Bailliet et al. (1994), N. O. Bianchi and F. Rothhammer (1995), and D. A. Merriwether (1994, 1995) suggest that multiple variants of each lineage entered the New World, and that additional unrelated lineages also entered. The researchers present the distribution of multiple variants of the four founding lineages, plus two additional lineages, which they call X6 and X7, throughout the New World, Siberia, and Asia. These distributions are strong evidence that at least nine different founding lineage haplotypes entered the New World. Further, they find these distributions among American Indians best fit a single wave of migration into the New World.

Genetic Materials and Lines
Francisco Rothhammer, Emoke Szathmary, and Ken Weiss provided the samples. They typed over 1300 American Indians for haplogroup-specific restriction sites and the 9-bp

region V deletion to examine the distribution of these haplotypes. From DNA extracted from blood or plasma the researchers typed Aymara, Atacameno, Yaghan, Pehuenche, and Huilliche Indians from Chile; Quechua Indians from Peru; Muskoke Indians from Oklahoma; Dogrib and Mohawk Indians from Canada; and Eskimos and Aleuts from Alaska.

Study

Mishmar, D., Ruiz-Pesini, E., Golik, P., Macaulay, V., Clark, A. G., Hosseini, S., Brandon, M., Easley, K., Chen, E. and Brown et, a. 2003. Natural selection shaped regional mtDNA variation in humans. Proceedings of The National Academy of Sciences of The United States of America, 100(1): 171-176.

Study Summary

Human mtDNA shows striking regional variation, traditionally attributed to genetic drift. However, it is not easy to account for the fact that only two mtDNA lineages (M and N) left Africa to colonize Eurasia and that lineages A, C, D, and G show a 5-fold enrichment from central Asia to Siberia. As an alternative to drift, natural selection might have enriched for certain mtDNA lineages as people migrated north into colder climates. To test this hypothesis the researchers analyzed 104 complete mtDNA sequences from all global regions and lineages. African mtDNA variation did not significantly deviate from the standard neutral model, but European, Asian, and Siberian plus American Indian variations did. Analysis of amino acid substitution mutations (nonsynonymous, Ka) versus neutral mutations (synonymous, Ks) (kaks) for all 13 mtDNA protein-coding genes revealed that the ATP6 gene had the highest amino acid sequence variation of any human mtDNA gene, even though ATP6 is one of the more conserved mtDNA proteins. Comparison of the kaks ratios for each mtDNA gene from the tropical, temperate, and arctic zones revealed that ATP6 was highly variable in the mtDNAs from the arctic zone, cytochrome b was particularly variable in the temperate zone, and cytochrome oxidase I was notably more variable in the tropics. Moreover, multiple amino acid changes found in ATP6, cytochrome b, and cytochrome oxidase I appeared to be functionally significant. From these analyses they concluded that selection may have played a role in shaping human regional mtDNA variation and that one of the selective influences was climate.

Genetic Materials and Lines

Fifty-six mtDNA sequences were available from the literature encompassing individuals sampled from African, European, and Asian populations based on their language groups and geographic distribution. The researchers analyzed 48 additional individuals from African, Asian, European, Siberian, and American Indian populations to complete a global survey of mtDNA variation.

Study

Monsalve, M. V., Helgason, A. and Devine, D. V. 1999. Languages, geography and HLA haplotypes in American Indian and Asian populations. Proceedings of the Royal Society of London Series B-Biological Sciences, 266(1434): 2209-2216.

Study Summary

A number of studies based on linguistic, dental and genetic data have proposed that the colonization of the New World took place in three separate waves of migration from North-East Asia. Recently, other studies have suggested that only one major migration occurred. It was the aim of this study to assess these opposing migration hypotheses using molecular-typed HLA class II alleles to compare the relationships between linguistic and genetic data in contemporary American Indian populations. The researcher's results suggest that gene flow and genetic drift have been important factors in shaping the genetic landscape of American Indian populations. They report significant correlations between genetic and geographical distances in American Indian and East Asian populations. In contrast, a less clear-cut relationship seems to exist between genetic distances and linguistic affiliation. In particular, the close genetic relationship of the neighboring Na-Dene

Athabaskans and Amerindian Salishans suggests that geography is the more important factor. Overall, their results are most congruent with the single migration model.

Genetic Materials and Lines

The researchers thank the First Nations People of British Columbia for their cooperation. Sheena Wilkie (Canadian Red Cross Society Blood Services) assisted in collecting the samples. Sofia Hashemi of the Canadian Red Cross, National Testing Laboratory, Ottawa, provided control samples.

Blood samples were collected from Salishans from the Soowahlie band as part of the recruitment of native people from British Columbia onto the Canadian Unrelated Bone Marrow Reistry (UBMDR). The participants completed a written questionnaire on linguistic, ethnic, and family heritage and consented in writing to the use of blood for the HLA analysis.

Study

Monsalve, M.V., Edin, G., and Devine, D.V. 1998. Analysis of HLA Class I and Class II in Na-Dene and Amerindian Populations from British Columbia, Canada. Human Immunology, 59(1): 48-55.

Study Summary

The researchers analyzed the distribution of HLA class II alleles and haplotypes in one Na-Dene (Athabaskan) group from British Columbia (Canada) by PCR amplification of the DRB1, DQA1 and DQB1 second exon sequences. They extended the typing of the DRB1 in an Amerindian group (Penutian) from British Columbia. The presence of the alleles DRB1* 0405, *0407 and *0410 only in Na-Dene and alleles DRB1*0408, *1301*1302, *1304, *1305, *1502 and *1601 only in Amerindians suggests separate origins of these two groups There were fifteen different DRB1/DQA1/DQB1 haplotypes. One unique haplotype previously reported in American Indians was found. Thirty-four per cent of Athabaskans presented American Indian haplotype

DRB1*1402/DQA1*0501/DQB1*0301. In addition, the results of this study are compatible with previous evidence with mitochondrial (mtDNA) polymorphisms indicating that Amerindians and Na-Dene populations derived from different migrations from Asia.

Genetic Materials and Lines
Not Available.

Study
Monsalve, M. V., Stone, A. C., Lewis, C. M., Rempel, A., Richards, M., Straathof, D. and Devine, D. V. 2002. Brief Communication: Molecular Analysis of the Kwaday Dan Ts'inchi Ancient Remains Found in a Glacier in Canada. American Journal of Physical Anthropology, 119: 288-291.

Study Summary
DNA was extracted from the frozen remains of a man found in the northwest corner of British Columbia, Canada, in 1999. His clothing was radiocarbon dated at ca. 550 years old. Nitrogen and carbon content in whole bone and collagen-type residue extracted from both bone and muscle indicated good preservation of proteinaceous macromolecules. Restriction enzyme analysis of mitochondrial DNA (mtDNA) determined that the remains belong to haplogroup A, one of the four major American Indian mtDNA haplogroups. Data obtained by PCR direct sequencing of the mtDNA control region, and by sequencing the clones from overlapping PCR products, were duplicated by an independent laboratory. Comparison of these mtDNA sequences with those of North American, Central American, South American, East Siberian, Greenlandic, and Northeast Asian populations indicates that the remains share an mtDNA type with North American, Central American, and South American populations.

Genetic Materials and Lines
The protocol and consent form for the study of ancient human remains were reviewed by the Clinical Research Ethics Board Committee at the University of British

Columbia. In August 1999, the frozen remains of a man were found on a glacier in Canada's Tatshenshini-Alsek Park near the Pacific coast border with Alaska. His clothing was radiocarbon-dated at 500 +/- 30 BP (Beattie et al., 2000). The local Champagne-Aishihik people named the remains "Kwaday Dan Ts' `nchi" (KDT), i.e., Long-Ago Person Found. With the consent of a committee comprised of First Nations representatives and British Columbia's Archaeology Branch, the researchers extracted mitochondrial DNA (mtDNA) from KDT hard and soft tissues and determined that his mtDNA belongs to haplogroup A, one of the four common American Indian haplogroups.

Study

Mori, M., Beatty, P. G., Graves, M., Boucher, K. M. and Milford, E. L. 1997. HLA gene and haplotype frequencies in the North American population - The National Marrow Donor Program Donor Registry. Transplantation, 64(7): 1017-1027.

Study Summary

As of May 1, 1995, the National Marrow Donor Program had a donor registry consisting of over 1.35 million HLA-typed volunteers recruited from most major cities and states in the United States. This registry represents the largest single HLA-typed pool of normal individuals in the world. The researchers analyzed the HLA-A, -B, -DR locus phenotypes of the National Marrow Donor Program donors in order to estimate gene and haplotype frequencies for major racial groups of the United States: Caucasian American, Asian American, African American, Latin American, and American Indian. The large size of the database allowed us to calculate the frequencies of relatively rare antigens and haplotypes with more accuracy than previous studies. They observed 89,522 distinguishable HLA-A, -B phenotypes in 1,351,260 HLA-A, -B-typed donors and 302,867 distinguishable HLA-A, -B, -DR phenotypes in 406,503 HLA-A, -B, -DR-typed donors. Gene and haplotype frequencies differed remarkably among the five racial groups, with African Americans and Asian Americans having a large number of haplotypes that were specific to their racial groups, whereas Caucasian Americans, Latin Americans, and American Indians shared a number of common haplotypes. These data represent an important resource for investigators in the fields of transplantation and population genetics. The gene and haplotype frequencies can be used to aid clinicians in advising patients about the probability of finding a match within a specific ethnic group, or to determine donor recruitment goals and strategies. The information is also a valuable resource for individuals who are interested in population genetics, selection and evolution of polymorphic human genes, and HLA-disease association.

Genetic Materials and Lines

Subjects were volunteers who had been recruited by the NMDP through its participating donor centers. Local histocompatibility laboratories phenotyped donors for HLA-A, -B, -DR antigens using standard microlymphocytotoxicity assays for A and B and serology or DNA typing for DR, which were reported at the serological level of resolution. HLA phenotypes were reported to the NMDP along with demographic information about the volunteer, including sex and race/ethnicity. Race is self-reported and broadly categorized as: Caucasian American, African American, Asian American, Latin American, and American Indian.

The study population consisted of 1,351,260 HLA-typed individuals, of whom 73.8% were Caucasian American, 8.1% African American, 6.0% Asian American, 7.4% Latin American, and 1.4% American Indian. Donors who did not have a stated ethnic group were excluded from the current analysis (3.2% of all donors). The subjects were recruited at donor centers distributed over most of the major cities and states. A total of 406,503 individuals, or 30% of the registered donors, were typed for HLA-A, -B, -DR antigens, and the remainder were typed only for the A and B locus antigens.

Study

O'Rourke, D. H., Hayes, M. G. and Carlyle, S. W. 2000. Spatial and temporal stability of mtDNA haplogroup frequencies in native North America. Human Biology, 72(1): 15-34.

Study Summary

The origin and maintenance of genetic variation in modern populations of North American aboriginal populations were investigated. A comparison of 6 ancient population samples of North America with 41 contemporary North American populations scattered throughout the continent confirmed earlier work that indicated substantial geographic restructuring of mtDNA haplogroup frequencies. The ancient samples displayed haplogroup profiles that were most similar to those of modern populations living in the same geographic regions today. This result suggests a surprising stability of mtDNA haplogroup profiles in indigenous populations of the Americas over the past 2,000 years plus of the Holocene.

Genetic Materials and Lines

The Northwestern Band of the Shoshone Nation gave permission to study the material from the Great Salt Lake Wetlands burials. The Aleut Corporation and the Chaluka Corporation gave permission to analyze the Aleut material. Anasazi material came from the American Museum of Natural History, New York, and the Aleut samples from the Smithsonian Institution, Washington, DC.

Study

O'Rourke, D. H., Mobarry, A. and Suarez, B. K. 1992. Patterns of Genetic Variation in Native America. Human Biology, 64(3): 417-434.

Study Summary

Allele frequencies from seven polymorphic red cell antigen loci (ABO, Rh, MN, S, P, Duffy, and Diego) were examined in 144 American Indian populations. Mean genetic distances (Nei's D) and the fixation index Fst are approximately equal for the North and South American samples but are reduced in the Central American geographic area. The relationship between genetic distance and geographic distance differs markedly across geographic areas. The correlation between geographic distance and genetic distance for the North and Central American data is twice as large as that observed for the South American samples. This geographic difference is confirmed in spatial autocorrelation analyses; no geographic structure is apparent in the South American data but geographic structure is prominent in North and Central American samples. These results confirm earlier observations regarding differences between North and South American gene frequency patterns.

Genetic Materials and Lines

The data came from seven red cell antigen loci used to construct genetic maps in earlier publications (Suarez, Crouse, and O'Rourke 1985; O'Rourke and Suarez 1985). In the North American samples 29 populations required the estimation of frequencies for a single system, whereas in the South American samples only 6 frequencies needed to be estimated.

Study

Paabo, S., Gifford, J.A., and Wilson, A.C. 1988. Mitochondrial DNA Sequences from a 7000-year-old Brain. Nucleic Acid Research, 16: 9775-9787.

Study Summary

Not Available.

Genetic Materials and Lines

Not Available.

Study

Parr, R.L., Carlyle, S.W., and O'Rourke, D.H. 1996. Ancient DNA Analysis of Fremont Amerindians of the Great Salt Lake Wetlands. American Journal of Physical Anthropology, 99(4): 507-518.

Study Summary

Skeletal remains of 47 individuals from the Great Salt Lake Wetlands, affiliated principally with Bear River (A.D. 400-1000) and Levee Phase (A.D. 1000-1350) Fremont cultural elements, were assessed for four mitochondrial DNA (mtDNA) markers that, in particular association, define four haplogroups (A, B, C, and D) widely shared among contemporary Amerindian groups. The most striking result is the absence of haplogroup A in this Fremont series, despite its predominance in contemporary Amerindian groups. Additionally, haplogroup B, defined by the presence of a 9bp deletion in region V, is present at the moderately high frequency of 60%. Haplogroups C and D are present at low frequencies. An additional haplotype, "N," observed in some modern populations and two other prehistoric samples, is also present in this Fremont skeletal collection.

Genetic Materials and Lines

Not Available.

Study

Pena, S.D., Santos, F.R., Bianchi, N.O., Bravi, C.M., Carnese, F.R., Rothhammer, F., et al. 1995. A Major Founder Y-Chromosome Haplotype in Amerindians. Nat Genet, 11(1): 15-16.

Study Summary

Not Available.

Genetic Materials and Lines

Not Available.

Study

Poinar, H. N., Kuch, M., Sobolik, K. D., Barnes, I., Stankiewicz, A. B., Kuder, T., Spaulding, W. G., Bryant, V. M., Cooper, A. and Paabo et, a. 2001. A molecular analysis of dietary diversity for three archaic American Indians. Proceedings Of The National Academy Of Sciences Of The United States Of America, 98(8): 4317-4322.

Study Summary

DNA was extracted from three fecal samples, more than 2,000 years old, from Hinds Cave, Texas. Amplification of human mtDNA sequences showed their affiliation with contemporary American Indians, while sequences from pronghorn antelope, bighorn sheep, and cottontail rabbit allowed these animals to be identified as part of the diet of these individuals. Furthermore, amplification of chloroplast DNA sequences identified eight different plants as dietary elements. These archaic humans consumed 2-4 different animal species and 4-8 different plant species during a short time period. The success rate for retrieval of DNA from paleofeces is in strong contrast to that from skeletal remains where the success rate is generally low. Thus, human paleofecal remains represent a source of ancient DNA that significantly complements and may in some cases be superior to that from skeletal tissue.

Genetic Materials and Lines
Not Available.

Study

Poloni, E. S., Semino, O., Passarino, G., Santachiara-Benerecetti, A. S., Dupanloup, I., Langaney, A. and Excoffier, L. 1997. Human Genetic Affinities for Y-Chromosome P49a,f/Taq1 Haplotypes Show Strong Correspondence with Linguistics. American Journal of Human Genetics, 61: 1015-1035.

Study Summary

Numerous population samples from around the world have been tested for Y chromosome-specific p49a,f/TaqI restriction polymorphisms. Here the researchers review the literature as well as unpublished data on Y-chromosome p49a,f/Taq haplotypes and provide a new nomenclature unifying the notations used by different laboratories. They use this large data set to study worldwide genetic variability of human populations for this paternally transmitted chromosome segment. The researchers observe, for the Y chromosome, an important level of population genetic structure among human populations (F_{ST}=.230, P<.001), mainly due to genetic differences among distinct linguistic groups of populations (F_{CT}=.246, P<.001). A multivariate analysis based on genetic distances between populations shows that human population structure inferred from the Y chromosome corresponds broadly in language families (r=.567, P<.001), in agreement with autosomal and mitochondrial data. Times of divergence of linguistic families, estimated from their internal level of genetic differentiation, are fairly concordant with current archaeological and linguistic hypotheses. Variability of the p49a,f/TaqI polymorphic marker is also significantly correlated with the geographic location of the populations (r=.613, P<.001), reflecting the fact that distinct linguistic groups generally also occupy distinct geographic areas. Comparison of Y-chromosome and mtDNA RFLPs in a restricted set of populations shows a globally high level of congruence, but it also allows identification of unequal maternal and paternal contributions to the gene pool of several populations.

Genetic Materials and Lines

The researchers selected samples from two distinct sources: 45 samples were selected from a total of 60 published samples, on the basis of sample size (n>20) and ethnic homogeneity; 13 additional samples taken from various Mediterranean populations have also been included, even though the data are still unpublished (A.S. Santachiara-Benerecetti, unpublished data). They thus gathered a total of 58 samples, representing the analysis of 3.767 chromosomes for the p49a,f/TaqI polymorphism. They used Mexican Indian sources previously published in Torroni et al. (1994)

Study

Reidla, M., Kivisild, T., Metspalu, E., Kaldma, K., Tambets, K., Tolk, H.-V., Parik, J., Loogvali, E.-L., Derenko, M. and Malyarchuk et, a. 2003. Origin and Diffusion of mtDNA Haplogroup X. American Journal Of Human Genetics, 73(6): 1178-1190.

Study Summary

A maximum parsimony tree of 21 complete mitochondrial DNA (mtDNA) sequences belonging to haplogroup X and the survey of the haplogroup-associated polymorphisms in 13,589 mtDNAs from Eurasia and Africa revealed that haplogroup X is subdivided into two major branches, here defined as "X1" and "X2." The first is restricted to the populations of North and East Africa and the Near East, whereas X2 encompasses all X mtDNAs from Europe, western and Central Asia, Siberia, and the great majority of the Near East, as well as some North African samples. Subhaplogroup X1 diversity indicates an early coalescence time, whereas X2 has apparently undergone a more recent population expansion in Eurasia, most likely around or after the last glacial maximum. It is notable that X2 includes the two complete American Indian X sequences that constitute the distinctive X2a clade, a clade that lacks close relatives in the entire Old World, including Siberia. The position of X2a in the phylogenetic tree suggests an early split from the other X2 clades, likely at the very beginning of their expansion and spread from the Near East.

Genetic Materials and Lines
Not Available.

Study

Rubicz, R., Melvin, K. L. and Crawford, M. H. 2002. Genetic evidence for the phylogenetic relationship between Na-Dene and Yeniseian speakers. Human Biology, 74(6): 743-760.

Study Summary

Ruhlen's hypothesis, based on linguistic evidence, for a common phylogenetic origin of Na-Dene and Yeniseian speakers is tested using genetic data. Gene frequency data for the Kets, the only surviving Yeniseian speakers, were collected during a field study in 1993. Data for several Na-Dene groups, as well as other New World and Siberian populations, were compiled from the literature. These data were analyzed using R-matrix, principal components analysis, and Mantel tests. In a comparison of 10 New World and Siberian populations using eight alleles, 55.8% of the variation was accounted for by the first principal component, and 22.1% of the variation was subsumed by the second principal component. Contrary to Ruhlen's interpretation of the linguistic data, analysis of the genetic data shows that the Na-Dene cluster with other American Indian populations, while the Kets genetically resemble the surrounding Siberian groups. This conclusion is further supported by correlations that are higher when the Kets are considered unrelated to Na-Dene speakers, and an insignificant partial correlation between genes and language when geography is held constant, indicating that spatial patterning accounts for most of the variation present in these populations.

Genetic Materials and Lines

Gene frequency data including blood groups, immunoglobulins, and mitochondrial DNA (mtDNA) haplogroups were collected from the literature for the following populations: Apache, Dogrib, Haida, Kutchin, Navajo, and Tlingit in the Na-Dene family; Blackfeet, Cree, Eskimo, Ojibwa, and Papago in non-Na-Dene of the Americas; and Altai, Asian Eskimo, Coastal Chukchi, Evenki, Forest Nentsi, Nganasan, Reindeer Chukchi, Sel'kup, and Yukaghir in Siberia. Blood samples were collected from the Kets by researchers from the University of Kansas. Blood typing was done at the Minneapolis War Memorial Blood Bank, and the immunoglobulins were characterized at the Analytical Genetic Testing Center, Denver, Colorado.

Study

Rubicz, R., Schurr, T. G., Babb, P. L. and Crawford, M. H. 2003. Mitochondrial DNA variation and the origins of the Aleuts. Human Biology, 75(6): 809-35.

Study Summary

The mitochondrial DNA (mtDNA) variation in 179 Aleuts from five different islands (Atka, Unalaska, Umnak, St. Paul, and St. George) and Anchorage was analyzed to better understand the origins of Aleuts and their role in the peopling of the Americas. Mitochondrial DNA samples were characterized using polymerase chain reaction amplification, restriction fragment length polymorphism analysis, and direct sequencing of the first hypervariable segment (HVS-I) of the control region. This study showed that Aleut mtDNAs belonged to two of the four haplogroups (A and D) common among American Indians. Haplogroup D occurred at a very high frequency in Aleuts, and this, along with their unique HVS-I sequences, distinguished them from Eskimos, Athapaskan Indians, and other northern Amerindian populations. While sharing several control region sequences (CIR11, CHU14, CIR60, and CIR61) with other circumarctic populations, Aleuts lacked haplogroup A mtDNAs having the 16265G mutation that are specific to Eskimo populations. R-matrix and median network analyses indicated that Aleuts were closest genetically to Chukotkan (Chukchi and Siberian Eskimos) rather than to American Indian or Kamchatkan populations (Koryaks and Itel'men). Dating of the Beringian branch of haplogroup A (16192T) suggested that populations ancestral to the Aleuts, Eskimos, and Athapaskan Indians emerged approximately 13,120 years ago, while Aleut-specific A and D sublineages were dated at 6539 +/- 3511 and 6035 +/- 2885 years, respectively. The researchers findings support the archaeologically based hypothesis that ancestral Aleuts crossed the Bering Land Bridge or Beringian platform and entered the Aleutian Islands from the east, rather than island hopping from Kamchatka into the western Aleutians. Furthermore, the Aleut migration most likely represents a separate event from those responsible for peopling the remainder of the Americas, meaning that the New World was colonized through multiple migrations.

Genetic Materials and Lines

The Aleut elder Alice Petrivelli and other Aleut individuals participated in this study. This study was approved by the University of Kansas Advisory Committee on Human Experimentation and several Aleut tribal organizations (Aleutian/Pribilof Island Association, Aleut Corporation, and tribal councils). Samples were collected from 179 Aleuts living in five, small Alaskan communities and one city. These included 17 from Atka, 17 from Umnak Island (Nikolski), and

37 from Unalaska (in the central and eastern Aleutians); 32 from St. George and 47 from St. Paul (in the Pribilofs); and 29 from Anchorage. Each participant was interviewed to obtain family background information (including birth locations), and to control for admixture by confirming Aleut ancestry on his or her maternal side. Cheek cell samples from the first 50 individuals were collected using OraSure kits (Analytical Genetic Testing Center, Denver, CO). The remaining samples were collected with sterile wooden applicators and stored in Tris-EDTA buffer.

Study

Ruiz-Linares, A., Ortiz-Barrientos, D., Figueroa, M., Mesa, N., Munera, J. G., Bedoya, G., Velez, I. D., Garcia, L. F., Perez-Lezaun, A., Bertranpetit, J., Feldman, M. W. and Goldstein, D. B. 1999. Microsatellites provide evidence for Y chromosome diversity among the founders of the New World. Proceedings of the National Academy of Sciences, 96(11): 6312-7.

Study Summary

Recently, Y chromosome markers have begun to be used to study American Indian origins. Available data have been interpreted as indicating that the colonizers of the New World carried a single founder haplotype. However, these early studies have been based on a few, mostly complex polymorphisms of insufficient resolution to determine whether observed diversity stems from admixture or diversity among the colonizers. Because the interpretation of Y chromosomal variation in the New World depends on founding diversity, it is important to develop marker systems with finer resolution. Here the researchers evaluate the hypothesis of a single-founder Y haplotype for Amerinds by using 11 Y-specific markers in five Colombian Amerind populations. Two of these markers (DYS271, DYS287) are reliable indicators of admixture and detected three non-Amerind chromosomes in the researchers sample. Two other markers (DYS199, M19) are single-nucleotide polymorphisms mostly restricted to American Indians. The relatedness of chromosomes defined by these two markers was evaluated by constructing haplotypes with seven microsatellite loci (DYS388 to 394). The microsatellite backgrounds found on the two haplogroups defined by marker DYS199 demonstrate the existence of at least two Amerind founder haplotypes, one of them (carrying allele DYS199 T) largely restricted to American Indians. The estimated age and distribution of these haplogroups places them among the founders of the New World.

Genetic Materials and Lines

The total number of unrelated male samples available for typing was 137 from five Colombian Amerind populations: 8 Embera, 10 Ingano, 40 Ticuna, 21 Wayuu (or Goajiro), 58 Zenu (or Sinu). The samples for the Zenu and Embera populations were available at Universidad de Antioquia and had been collected for other studies. In three instances (Ticuna, Wayuu, and Ingano), samples were collected from informed consenting individuals.

Study

Saillard, J., Forster, P., Lynnerup, N., Bandelt, H.-J., and Norby, S. 2000. mtDNA Variation Among Greenland Eskimos: The Edge of the Beringian Expansion. American Journal of Human Genetics, 67, 718-726.

Study Summary
Not Available.

Genetic Materials and Lines
Not Available.

Study

Santos, F. R., Pandya, A., Tyler-Smith, C., Pena, S. D. J., Schanfield, M., Leonard, W. R., Osipova, L., Crawford, M. H. and Mitchell, R. J. 1999. The Central Siberian Origin for American Indian Y Chromosomes. American Journal of Human Genetics, 64: 619-628.

Study Summary

Y chromosomal DNA polymorphisms were used to investigate Pleistocene male migrations to the American continent. In a worldwide sample of 306 men, the researchers obtained 32 haplotypes constructed with the variation found in 30 distinct polymorphic sites. The major Y haplotype present in most American Indians was traced back to recent ancestors common with Siberians, namely, the Kets and Altaians from the Yenissey River Basin and Altai Mountains, respectively. Going further back, the next common ancestor gave rise also to Caucasoid Y chromosomes, probably from the central Eurasian region. This study, therefore, suggests a predominantly central Siberian origin for American Indian paternal lineages for those who could have migrated to the Americas during the Upper Pleistocene.

Genetic Materials and Lines

Most of the 306 male samples were obtained as DNA or were extracted from plugs prepared for pulsed-field gel electrophoresis. Samples from Europeans (most were British), Indians (India and Sri Lanka), Africans (Kenyans, Pygmies, and San), central East Asians (Chinese and Japanese), Mongolians (Khalkhs), and Siberians (Buryats, Yakuts, Evenkis, Altaians, and Kets) were subsets of those described elsewhere (Mathias et al. 1994; Zerjal et al. 1997). Ten samples, from south and central Amerindians and a Na-Dene, were purchased from the National Institute of General Medical Science, and an additional 10 American Indian samples (not Aleut-Eskimos) came from paternity tests in North America.

Study

Santos, F.R., Hutz, M.H., Coimbra, C.E.A., Santos, R.V., Salzano, F.M., and Peena, S.D.J. 1995. Further Evidence for the Existence of a Major Founder Y Chromosome Haplotype in Amerindians. Brazilian Journal of Genetics, 18: 669-672.

Study Summary
    Not Available.

Genetic Materials and Lines
    Not Available.

Study
    Santos, F.R., Rodriguez-Delfin, L., Pena, S.D., Moore, J., and Weiss, K.M. 1996. North and South Amerindians may have the same Major Founder Y Chromosome Haplotype. American Journal of Human Genetics, 58(6): 1369-1370.

Study Summary
    Not Available.

Genetic Materials and Lines
    Not Available.

Study
    Schurr, T., Ballinger, S., Gan, Y.-Y., Hodge, J. A., Weiss, K. M. and Wallace, D. 1990. Amerindian Mitochondrial DNAs Have Rare Asian Mutations at High Frequencies, Suggesting They Derived from Four Primary Maternal Lineages. American Journal of Human Genetics, 46: 613-623.

Study Summary
    The mitochondrial DNA (mtDNA) sequence variation of the South American Ticuna, the Central American Maya, and the North American Pima was analyzed by restriction-endonuclease digestion and oligonucleotide hybridization. The analysis revealed that Amerindian populations have high frequencies of mtDNAs containing the rare Asian RFLP HincII morph 6, a rare HaeIII site gain, and a unique AluI site gain. In addition, the Asian-specific deletion between the cytochrome c oxidase subunit II (COII) and tRNAlys genes was also prevalent in both the Pima and the Maya. These data suggest that Amerindian mtDNAs derived from at least four

primary maternal lineages, that new tribal-specific variants accumulated as these mtDNAs became distributed throughout the Americas, and that some genetic variation may have been lost when the progenitors of the Ticuna separated from the North and Central American populations.

Genetic Materials and Lines

F.M. Salzano (Department de Genetica, Instituto de Biociencias, Universidade Federal do Rio Grande do Sul, Porto Alegre, Brazil); J.V. Neel (Department of Human Genetics, University of Michigan Medical School); Fundacao National do Indio (FUNAI); Instituto Nacional de Pesquisas da Amazonia (INPA); W.F. and J.A. Bodmer (Imperial Cancer Research Fund Laboratories, London) collected and preserved Ticuna blood samples; the National Science Foundation make available the facilities of the research vessel Alpha Helix during July and August of 1976; and J.R. Kidd (Department of Human Genetics, Yale University School of Medicine) prepared the Maya cell lines and DNAs. The Gila River Community and Drs. D.J. Pettitt and M.J. Carraher, as well as the laboratory and field staff of the Diabetes and Arthritis Epidemiology Section, NIDDK, obtained Pima blood samples and demographic data.

Study

Schurr, T.G., and Sherry, S.T. 2004. Mitochondrial DNA and Y Chromosome Diversity and the Peopling of the Americas: Evolutionary and Demographic Evidence. American Journal of Human Biology, 16: 420-439.

Study Summary

A number of important insights into the peopling of the New World have been gained through molecular genetic studies of Siberian and American Indian populations. While there is no complete agreement on the interpretation of the mitochondrial DNA (mtDNA) and Y Chromosome (NRY) data from these groups, several generalizations can be made. To begin with, the primary migration of ancestral Asians expanded from south-central Siberia into the Americas gave rise to ancestral Amerindians. The initial migration seems to have occurred between 20,000-15,000 ybp, before the emergence of Clovis lithic sites in North America. Because an interior route through northern North America was unavailable for human passage until 12,550 ybp, after the last glacial maximum, these ancestral groups must have used a coastal route to reach South America by 14,675 ybp, the date of the Monte Verde site in southern Chile. The initial migration appears to have brought mtDNA haplogroups A-D and NRY haplogroups P-M45a and Q-242/Q-M3 to the Americas, with these genetic lineages becoming widespread in the Americas. A second expansion that perhaps coincided with the opening of the ice-free corridor probably brought mtDNA haplogroup X and NRY haplogroups P-M45b, C-M130, and R1a1-M17 to North and Central America. Finally, populations that formerly inhabited Beringia expanded into northern North America after the last glacial maximum and gave rise to Eskimo-Aleuts and Na-Dene Indians.

Genetic Materials and Lines
Not Available.

Study
Schurr, T. G., Sukernik, R. I., Starikovskaya, Y. B. and Wallace, D. C. 1999. Mitochondrial DNA Variation in Koryaks and Itel'men: Population Replacement in the Okhotsk Sea-Bering Sea Region During the Neolithic. American Journal of Physical Anthropology, 108: 1-39.

Study Summary
In this study, the researchers analyzed the mitochondrial DNA (mtDNA) variation in 202 individuals representing one Itel'men and three Koryak populations from different parts of the Kamchatka peninsula. All mtDNAs were subjected to high resolution restriction (RFLP) analysis and control (CR) sequencing, and the resulting data were combined with those available from other Siberian and east Asian populations and subjected to statistical and phylogenetic analysis. Together,

the Koryaks and Itel'men were found to have mtDNAs belonging to three (A,C, and D) of the four major haplotype groups (haplogroups) observed in Siberian and American Indian populations (A-D). In addition, they exhibited mtDNAs belonging to haplogroups G, Y, and Z, which were formerly called "other" mtDNAs. While Kamchatka harbored the highest frequencies of haplogroup G mtDNAs, which were widely distributed in eastern Siberian and adjacent east Asian populations, the distribution of haplogroup Y was restricted within a relatively small area and pointed to the lower Amur River-Sakhalin Island region as its place of origin. In contrast, the pattern of distribution and the origin of haplogroup Z mtDNAs remained unclear. Furthermore, phylogenetic and statistical analyses showed that Koryaks and Itel'men had stronger genetic affinities with eastern Siberian/east Asian populations than to those of the northern Pacific Rim. These results were consistent with colonization events associated with the relatively recent immigration to Kamchatka of new tribes from the Siberian mainland region, although remnants of ancient Beringian populations were still evident in the Koryak and Itel'men gene pools.

Genetic Materials and Lines

Hospital staff and doctors in the villages of Ossora, Karaga, and Tymlat assisted with this project and the Koryak people from these villages participated in this research. The hospital staff in the villages of Voyampolka and Kovran also assisted with this project and the Koryak and Itel'men people from these villages participated in this research.

In July–August 1993, blood samples were collected from 104 Koryaks residing in three geographically proximate villages, Karaga, Ossora, and Tymlat, located in the Karaginskiy District of the Koryak Autonomous Region. Almost all of the Koryaks living in Ossora and Tymlat represented the Aluitor subgroup. Of this total, 27 and 12 individuals were born in the villages of Rekinniki and Anapka, respectively, both of which were closed in the early 1960s. Once considered to speak a distinct language, the Aluitor were the largest territorial group of Koryaks and occupied the whole Kamchatka isthmus and adjacent Bering Sea coast, combining small-scale reindeer herding with sea mammal hunting and fishing. Whereas approximately half of the Karaginskiy District sample consisted of Aluitor Koryaks, the remaining half represented the Karagin Koryaks. Speaking a slightly modified dialect of the Aluitor language, the Karagin Koryaks traditionally occupied the territory south of Tymlat, including Karaginskiy Island and settlements scattered along the Bering Sea coast. In June 1996, blood samples were obtained from 51 Koryak and 47 Itel'men residing in the villages of Voyampolka and Kovran, which are located in the Tigil'skiy District of the Koryak Autonomous Region. The Voyampolka sample was comprised of persons having mixed Maritime and Reindeer Koryak origins but who could be considered to belong to the Palan subgroup. The Palan Koryaks traditionally lived in several settlements scattered along the Okhotsk Sea coast of the peninsula between Voyampolka in the south and Lesnaya in the north but now reside mostly in these two villages. The Itel'men samples were obtained from individuals living in Kovran. All of these persons were born in or derived from one of a number of traditional settlements scattered along the Okhotsk Sea coast between Sopochnoye in the south and Tigil' in the north, including the villages of Kovran, Napana, Utkholok, Moroshechnoye, Belogolovoye, and Verkhneye Kharyuzovo. With the exception of Kovran and Verkhneye Kharyuzovo, none of these former Itel'men villages exist today. Blood samples were collected from each participant with informed consent in two sets of 10 ml ACD anti-coagulant tubes and kept refrigerated in the local hospitals until shipped or brought back to Atlanta for molecular genetic analysis. All individuals who participated in these studies were interviewed about their family histories, which in turn were verified by senior members of the community for accuracy. Only those persons who lacked maternal and paternal Russian or nonrelated ancestry through three generations were selected for the collection of blood samples, although samples were also obtained from four Evens who were the marital partners of Koryak participants. Based on these genealogical data, the researchers estimated that approximately half of the Koryaks and most of the Itel'men are of mixed Russian- Koryak or Russian-Itel'men ancestry, respectively, and consider themselves Koryak or Itel'men by nationality primarily because of their maternal ancestry.

Study

Schurr, T. G. and Wallace, D. C. 2002. Mitochondrial DNA Diversity in Southeast Asian Populations. Human Biology, 74(3): 431-52.

Study Summary

In a previous study of Southeast Asian genetic variation, these researchers characterized mitochondrial DNAs (mtDNAs) from six populations through high-resolution restriction fragment length polymorphism (RFLP) analysis. Their analysis revealed that these Southeast Asian populations were genetically similar to each other, suggesting they had a common origin. However, other patterns of population associations also emerged. Haplotypes from a major founding haplogroup in Papua New Guinea were present in Malaysia; the Vietnamese and Malaysian aborigines (Orang Asli) had high frequencies of haplogroup F, which was also seen in most other Southeast Asian populations; and haplogroup B, defined by the Region V 9-base-pair deletion, was present throughout the region. In addition, the Malaysian and Sabah (Borneo) aborigine populations exhibited a number of unique mtDNA clusters that were not observed in other populations. Unfortunately, it has been difficult to compare these patterns of genetic diversity with those shown in subsequent studies of mtDNA variation in Southeast Asian populations because the latter have typically sequenced the first hypervariable segment (HVS-I) of the control region (CR) sequencing rather than used RFLP haplotyping to characterize the mtDNAs present in them. For this reason, the researchers sequenced the HVS-I of Southeast Asian mtDNAs that had previously been subjected to RFLP analysis, and compared the resulting data with published information from other Southeast Asian and Oceanic groups. The researchers findings reveal broad patterns of mtDNA haplogroup distribution in Southeast Asia that may reflect different population expansion events in this region over the past 50,000-5000 years.

Genetic Materials and Lines

The data discussed in this study derive mostly from published sources. The RFLP haplotype data come from Horai et al. (1984), Horai and Matsunaga (1986), Ballinger et al. (1992), Passarino et al. (1993), Torroni et al. (1993a, 1994a), Kolman et al. (1996), and Kivisild et al. (1999). The HVS-I sequence data come mostly from Torroni et al. (1993a), Lum et al. (1994), Melton et al. (1995, 1998), Redd et al. (1995), Horai et al. (1996), Kolman et al. (1996), Lum and Cann (1998), Redd and Stoneking (1999), and Yao et al. (2000). The unpublished HVS-I sequence data for Nepalese groups were first reported in Schurr et al. (2000), while the preliminary HVS-I sequence data for the Bornean, Malaysian, and Vietnamese samples were first reported in Wallace and Schurr (2001). All of these HVS-I sequences were determined by standard methods (Schurr et al. 1999). Nearly all of these samples share one thing in common: they have been screened for the Region V 9-bp deletion that characterizes haplogroup B (Ballinger et al. 1992; Torroni et al. 1992). Among the papers from which deletion frequencies were obtained are Horai and Matsunaga (1986), Cann et al. (1987), Harihara et al. (1988), Hertzberg et al. (1989), Stoneking et al. (1990), Ballinger et al. (1992), Passarino et al. (1993), Torroni et al. (1993a, 1994a), Lum et al. (1994), Melton et al. (1995, 1998), Redd et al. (1995), Betty et al. (1996), Kolman et al. (1996), and Lum and Cann (1998).

Study

Scozzari, R., Cruciani, F., Santolamazza, P., Sellitto, D., Cole, D.E., Rubin, L.A. et al. 1997. mtDNA and Y Chromosome-specific Polymorphisms in Modern Ojibwa: Implications About the Origin of Their Gene Pool. American Journal of Human Genetics, 60(1): 241-244.

Study Summary

Not Available.

Genetic Materials and Lines
Not Available.

Study
Seielstad, M., Bekele, E., Ibrahim, M., Toure, A. and Traore, M. 1999. A View of Modern Human Origins from Y Chromosome Microsatellite Variation. Genome Research, 9: 558-567.

Study Summary
The idea that all modern humans share a recent (within the last 150,000 years) African origin has been proposed and supported on the basis of three observations. Most genetic loci examined to date have 1) shown greater diversity in African populations than in others, 2) placed the first branch between African and all non-African populations in phylogenetic trees, and 3) indicated recent dates for either the molecular coalescence (with the exception of some autosomal and X-chromosomal loci) or for the time of separation between African and non-African populations. The researches analyzed variation at 10 Y chromosome microsatellite loci that were typed in 506 males representing 49 populations and every inhabited continent and find significantly greater Y chromosome diversity in Africa than elsewhere, find the first branch in phylogenetic trees of the continental populations to fall between African and all non-African populations, and date this branching with the distance measure to 5800-17,400 or 12,800-36,800 years BP depending on the mutation rate used. The magnitude of the excess Y chromosome diversity in African populations appears to result from a greater antiquity of African populations rather than a greater long-term effective population size. These observations are most consistent with a recent African origin for all modern humans.

Genetic Materials and Lines
Trefor Jenkins and S. Qasim Mehdi provided for DNA samples from South African and Pakistani populations, respectively.

Study

Shields, G. F., Schmiechen, A. M., Frazier, B. L., Redd, A., Voevoda, M. I., Reed, J. K. and Ward, R. H. 1993. mtDNA sequences suggest a recent evolutionary divergence for Beringian and northern North American populations. American Journal Of Human Genetics, 53(3): 549-562.

Study Summary

Conventional descriptions of the pattern and process of human entry into the New World from Asia are incomplete and controversial. In order to gain an evolutionary insight into this process, the researchers have sequenced the control region of mtDNA in samples of contemporary tribal populations of eastern Siberia, Alaska, and Greenland and have compared them with those of Amerind speakers of the Pacific Northwest and with those of the Altai of central Siberia. Specifically, the researchers have analyzed sequence diversity in 33 mitochondrial lineages identified in 90 individuals belonging to five Circumpolar populations of Beringia, North America, and Greenland: Chukchi from Siberia, Inupiaq Eskimos and Athapaskans from Alaska, Eskimos from West Greenland, and Haida from Canada. Hereafter, the researchers refer to these five populations as "Circumarctic peoples." These data were then compared with the sequence diversity in 47 mitochondrial lineages identified in a sample of 145 individuals from three Amerind-speaking tribes (Bella Coola, Nuu-Chah-Nulth, and Yakima) of the Pacific Northwest, plus 16 mitochondrial lineages identified in a sample of 17 Altai from central Siberia. Sequence diversity within and among Circumarctic populations is considerably less than the sequence diversity observed within and among the three Amerind tribes. The similarity of sequences found among the geographically dispersed Circumarctic groups, plus the small values of mean pairwise sequence differences within Circumarctic populations, suggest a recent and rapid evolutionary radiation of these populations. In addition, Circumarctic populations lack the 9-bp deletion, which has been used to trace various migrations out of Asia, while populations of southeastern Siberia possess this deletion. On the basis of these observations, while the evolutionary affinities of American Indians extend west to the Circumarctic populations of eastern Siberia, they do not include the Altai of central Siberia.

Genetic Materials and Lines

Dr. Ralph Wells of Fairbanks Memorial Hospital helped coordinate the collection of placental tissues. T.I. Astakchova and V.L. Feigin helped coordinate collection of tissues from the Beringian regions of Siberia. Mark Lathrop, Chris Salski, and Catherine Enel collected the samples from West Greenland Eskimo and Gerry Nepom collected the Yakima samples.

mtDNA was purified from whole blood of 17 Altai from the villages of Ust-Kan (4 individuals), Ulagan (6 individuals), and Chibit (7 individuals) of the former Gorno-Altai Autonomous Territory, Siberia; 7 Chukchi from the villages of Ust-Belaia (4 individuals), Kanchalan (2 individuals), and Uelen (1 individual) of the former Chukchi Autonomous Territory, Chukotka, and 6 Siberian Eskimos from the villages of New Chaplino (5 individuals) and Lorino (1 individual) of Chukotka Peninsula. DNA was isolated by cesium-choloride density-gradient centrifugation from placenta of 5 Inupiaq mothers and 21 Athapaskan mothers who gave birth to their children at Fairbanks Memorial Hospital. The 5 Inupiaq Eskimos came from the villages of Shaktoolik (1 individual), and Tanana (2 individuals), while the 21 Athapaskans came from nine communities in interior Alaska: Tanana (6 individuals), Fairbanks (5 individuals), Hughes (2 individuals), Fort Yukon (3 individuals), with one individual each coming from St. Mary's, Nulato, Ruby, Koyukuk, and Stevens Village. DNA was also extracted from serum samples of 17 Eskimos from West Greenland and from whole blood samples of 42 Yakima from the Yakima Indian Reservation in Washington State.

Study

Silva, W. A., Jr., Bonatto, S. L., Holanda, A. J., Ribeiro-Dos-Santos, A. K., Paixao, B. M., Goldman, G. H., Abe-Sandes, K., Rodriguez-Delfin, L. and Barbosa et, a. 2002. Mitochondrial genome diversity of American Indians supports a single early entry of founder populations into America. American Journal Of Human Genetics, 71(1): 187-192.

Study Summary

There is general agreement that the American Indian founder populations migrated from Asia into America through Beringia sometime during the Pleistocene, but the hypotheses concerning the ages and the number of these migrations and the size of the ancestral populations are surrounded by controversy. DNA sequence variations of several regions of the genome of American Indians, especially in the mitochondrial DNA (mtDNA) control region, have been studied as a tool to help answer these questions. However, the small number of nucleotides studied and the non-clocklike rate of mtDNA control-region evolution impose several limitations to these results. Here the researchers provide the sequence analysis of a continuous region of 8.8 kb of the mtDNA outside the D-loop for 40 individuals, 30 of whom are American Indians whose mtDNA belongs to the four founder haplogroups. Haplogroups A, B, and C form monophyletic clades, but the five haplogroup D sequences have unstable positions and usually do not group together. The high degree of similarity in the nucleotide diversity and time of differentiation (i.e., approximately 21,000 years before present) of these four haplogroups support a common origin for these sequences and suggest that the populations who harbor them may also have a common history. Additional evidence supports the idea that this age of differentiation coincides with the process of colonization of the New World and supports the hypothesis of a single and early entry of the ancestral Asian population into the Americas.

Genetic Materials and Lines

In the present study, the researchers analyzed an 8,829-nt segment of the mitochondrial genome of 40 individuals, most of them American Indians. Since the analysis of the data, together with those from Ingman et al. (2000) for ~57% of the mtDNA molecule demonstrated that the results are similar to those obtained by sequencing the whole molecule, they restricted their analysis to a segment of 8.8 kb of mtDNA. The segment sequenced extends from nt 7148 to nt 15976 in the reference sequence J01415 (GenBank). A total of 30 American Indians of different linguistic stock were selected, to include a well-balanced representation of the four haplogroups based on the HVSI region and RFLP: Yanomama, Arara, Waiampi, Tyrio, Poturujara, Katuena, Kayapo, and Guarani (all from Brazil); and five Quechua from Peru. In addition, the following non–American Indians were also included: four African Brazilians, three Brazilians of Japanese origin (Asian Brazilians), and three white Brazilians.

Study

Smith, D. G., Malhi, R., Eshleman, J., Lorenz, J. G. and Kaestle, F. A. 1999. Distribution of mtDNA Haplogroup X Among Native North Americans. American Journal of Physical Anthropology, 110: 271-284.

Study Summary

Mitochondrial DNA samples of 70 American Indians, most of whom had been found not to belong to any of the four common American Indian haplogroups (A, B, C, D), were analyzed for the presence of Dde I site losses at np 1715 and np 10394. These two mutations are characteristic of haplogroup X, which might be of European origin. The first hypervariable segment (HVSI) of the non-coding control region (CR) of mtDNA of a representative selection of samples exhibiting these mutations was sequenced to confirm their assignment to haplogroup X. Thirty-two of the samples exhibited the restriction site losses characteristic of haplogroup X and, when sequenced, a representative selection (n=11) of these exhibited the CR mutations commonly associated with haplogroup X, C-T transitions at np 16278 and 16223, in addition to as many as three other HVSI mutations. The wide distribution of this haplogroup throughout North America, and its prehistoric presence there, are consistent with its being a fifth founding haplogroup exhibited by about 3% of modern American Indians. Its markedly nonrandom distribution with high frequency in certain regions, as for the other four major mtDNA haplogroups, should facilitate establishing ancestor/descendant relationships between modern and prehistoric groups of American Indians. The low frequency of haplogroups other than A, B, C, D, and X among the samples studied suggests a paucity of both recent non-American Indian maternal admixture in alleged fullblood American Indians and mutations at the restriction sites that characterize the five haplogroups as well as the absence of additional (undiscovered) founding haplogroups.

Genetic Materials and Lines

The Indian Health Service Facilities provided the samples studied. The researchers have previously characterized the geographic and ethnic distributions of haplogroups A, B, C, and D in a sample of 829 Native North Americans of alleged unmixed ancestry (Lorenz and Smith, 1996). Twenty additional samples of sera from Chippewa from the Lac Courte Oreilles community in Hayward, Wisconsin (or Southern Ojibwa) that were determined to be members of "other" haplogroups as parts of unrelated studies, and one sample each representing the Pomo, Blackfoot, and Sioux were also included in the present study. A total of 30 individuals from a sample of 189 Salteaux (or Northwestern) Chippewa, not previously screened for haplogroups A, B, C or D, were screened only for the presence of the Dde I site losses at np 1715 and 10394 that characterize haplogroup X. Since samples from other Chippewa groups had been found to exhibit this mutation, they studied 30 additional samples from a Salteaux Chippewa group merely to confirm its presence. Finally, their screening of a sample of 118 additional Northern Paiute samples yielded no members of haplogroups other than A, B, C, or D.

Study

Starikovskaya, Y. B., Sukernik, R. I., Schurr, T. G., Kogelnik, A. M. and Wallace, D. C. 1998. mtDNA diversity in Chukchi and Siberian Eskimos: Implications for the genetic history of ancient Beringia and the peopling of the New World. American Journal of Human Genetics, 63(5): 1473-1491.

Study Summary

The mtDNAs of 145 individuals representing the aboriginal populations of Chukotka- the Chukchi and Siberian Eskimos- were subjected to RFLP analysis and control-region sequencing. This analysis showed that the core of the genetic makeup of the Chukchi and Siberian Eskimos consisted of three (A, C, and D) of the four primary mtDNA haplotype groups (haplogroups) (A-D) observed in American Indians, with haplogroup A being the most prevalent in both Chukotkan populations. Two unique haplotypes belonging to haplogroup G (formerly called "other" mtDNAs) were also observed in a, few Chukchi, and these have apparently been acquired through gene flow from adjacent Kamchatka, where haplogroup G is prevalent in the Koryak and Itel'men. In addition, a 16111CàT transition appears to delineate an "American" enclave of haplogroup A mtDNAs in northeastern Siberia, whereas the 16192CàT transition demarcates a "northern Pacific Rim" cluster within this haplogroup. Furthermore, the sequence-divergence estimates for haplogroups A, C, and D of Siberian and American Indian populations indicate that the earliest inhabitants of Beringia possessed a limited number of founding mtDNA haplotypes and that the first humans expanded into the New World similar to 34,000 years before present (YBP). Subsequent migration 16,000-13,000 YBP apparently brought a restricted number of haplogroup B haplotypes to the Americas. For millennia, Beringia may have been the repository of the respective founding sequences that selectively penetrated into northern North America from western Alaska.

Genetic Materials and Lines

Dr. Vladimir V. Slavinskiy (District Hospital, Provideniya, Chukotkan Autonomous Region), helped with the fieldwork in Chukotka, and the Chukchi and Siberian Eskimo people openly participated in this research.

In October 1994 and April 1995, blood samples from 66 Chukchi and 79 Siberian Eskimos were collected in New Chaplino, Sireniki, Provideniya, and Anadyr, of the Chukotkan Autonomous Territory. Genealogical information was used to select unadmixed, unrelated individuals as potential blood donors. After informed consent was obtained, blood specimens were collected from participants, by venipuncture, into two sets of 10-ml ACD Vacutainer tubes (Becton-Dickinson). These samples were shipped to Emory University for subsequent molecular-genetic analysis. Of the entire Chukchi sample drawn from New Chaplino, Provideniya, Sireniki, and Anadyr, 55 individuals either had been born or had derived from the coastal villages of Neshkan, Enurmino, Inchoun, Uelen, Lorino, Yanrakinnot, Nunligran, and Enmelen and adjacent tundra camps (Kurupka) of the Chukchi peninsula, whereas 11 Chukchi either had been born or had derived from the Amguyema/Anadyr River area of interior Chukotka, the homeland of the reindeer Chukchi. In fact, the Eskimo sample assigned to the Chaplin tribe consists of individuals who either had been born or had derived from at least six tiny villages that had been scattered along the southeastern tip of the peninsula before being abandoned or closed during the 1940s and 1950s. Similarly, the Sireniki sample consisted of the Eskimos who had been born in Sireniki and a few other nearby but no longer existing villages. The Naukan Eskimos were represented by several individuals presently living in Sireniki, Provideniya, and New Chaplino.

Study

Steadman, D. W. 2001. Mississippians in Motion? A Population Genetic Analysis of Interregional Gene Flow in West-Central Illinois. American Journal of Physical Anthropology, 114: 61-73.

Study Summary

Population genetic and biological distance studies of Late Woodland and Mississippian populations from west-central Illinois have provided insight into a number of prehistoric demographic processes at the regional level. However, a formal analysis of diachronic interregional gene flow has not been attempted within a population genetics framework. In this study, cranial measurements of 489 individuals from 13 skeletal samples across the central and lower Illinois valleys are analyzed to address two central issues. First, the potential impact of Cahokia's decline and associated demographic events on the population structure of west-central Illinois Mississippians is examined. Second, the Mississippian and Late Woodland interregional migration patterns are compared to determine if geographic and/or cultural boundaries affected local population structure. Following Relethford and Blangero (1990), R matrix methods are utilized to calculate observed and expected phenotypic variances, minimum genetic distances, and FST values in order to detect patterns of differential external gene flow over time. The results indicate that Late Woodland peoples had a larger sphere of biological interaction than Mississippians. In the Mississippian period, culturally imposed barriers paralleled geographic boundaries between regions such that the geographic distribution of biological variation closely adheres to a classic isolation-by-distance model. Further, intraregional population movement was a more significant contributor to Mississippian population structure than interregional gene flow, even during periods of sociopolitical strife. Small-scale intraregional shuffling is consistent with other recent studies of prehistoric Mississippian biocultural and geographic landscapes in the southeast United States.

Genetic Materials and Lines

Della Cook, Duane Esarey, Alan Harn, George Milner, and Larry Conrad facilitated access to the skeletal materials. A total of 489 adults of both sexes from 13 skeletal samples of the central and lower Illinois valleys is included in the analyses. The Dickson Mounds sample is divided into two cultural phases, Eveland and Larson, based on burial goods, burial position, and mound affiliation. The Eveland and Larson phase samples from Dickson Mounds will be referred to in the analyses as discrete samples. A suite of 48 measurements was originally recorded for each individual based on a set of standard measurements derived in large part from North American

American Indian populations. Following analyses of intertrait correlations and intra and inter observer errors, the latter with Droessler (1981) who compiled the lower Illinois valley data, the suite is reduced to 11 variables: maximum length, frontal chord, occipital chord, biauricular breadth, maximum malar length, nasal height, minimum frontal breadth, midfacial breadth, biasterionic breadth, minimum ramus breadth, and maximum mandibular condyle length.

Study
Stone, A.C., and Stoneking, M. 1993. Ancient DNA from a pre-Columbian Amerindian Population. American Journal of Physical Anthropology, 92: 463-471.

Study Summary
Not Available.

Genetic Materials and Lines
Not Available.

Study
Stone, A.C., and Stoneking, M. 1998. mtDNA Analysis of a Prehistoric Oneota Population: Implications for the Peopling of the New World. American Journal of Human Genetics, 62: 1153-1170.

Study Summary
mtDNA was successfully extracted from 108 individuals from the Norris Farms Oneota, a prehistoric Native American population, to compare the mtDNA diversity from a pre-Columbian population with contemporary American Indian and Asian mtDNA lineages and to examine hypotheses about the peopling of the New World. Haplogroup and hypervariable region I sequence data indicate that the lineages from haplogroup A, B, C, and D are the most common among Native Americans but that they were not the only lineages brought into the Americas from Asia. The mtDNA evidence does not support the three-wave hypothesis of migration into

the Americas but rather suggests a single "wave" of people with considerable mtDNA diversity that exhibits a signature of expansion 23,000-37,000 years ago.

Genetic Materials and Lines
Not Available.

Study
Takahashi, N., Takahashi, Y., Blumberg, B. S. and Putnam, F. W. 1987. Amino Acid Substitutions in Genetic Variants of Human Serum Albumin and in Sequences Inferred from Molecular Cloning. Proceedings of the National Academy of Sciences, 84: 4413-4417.

Study Summary
The structural changes in four genetic variants of human serum albumin were analyzed by tandem high-pressure liquid chromatography (HPLC) of the tryptic peptides, HPLC mapping and isoelectric focusing of the CNBr fragments, and amino acid sequence analysis of the purified peptides. Lysine-372 of normal (common) albumin A was changed to glutamic acid both in albumin Naskapi, a widespread polymorphic variant of North American Indians, and in albumin Mersin found in Eti Turks. The two variants also exhibited anomalous migration in NaDodSO4/PAGE, which is attributed to a conformational change. The identity of albumins Naskapi and Mersin may have originated through descent from a common mid-Asiatic founder of the two migrating ethnic groups, or it may represent identical but independent mutations of the albumin gene. In albumin Adana, from Eti Turks, the substitution site was not identified but was localized to the region from positions 447 through 548. The substitution of aspartic acid-550 by glycine was found in albumin Mexico-2 from four individuals of the Pima tribe. Although only single point substitutions have been found in these and in certain other genetic variants of human albumin, five differences exist in the amino acid sequences inferred from cDNA sequences by workers in three other laboratories. However, their results on albumin A and on 14 different genetic variants accord with the amino acid

sequence of albumin deduced from the genomic sequence. The apparent amino acid substitutions inferred from comparison of individual cDNA sequences probably reflect artifacts in cloning or in cDNA sequence analysis rather than polymorphism of the coding sections of the albumin gene.

Genetic Materials and Lines

Dr. T. Isobe provided data on albumins B, Yanomama, and other variants. Dr. J. V. Neel provided reference specimens of albumins Yanomama, Makiritare-1, and Maku. Dr. C. Satoh (Hiroshima, Japan) provided albumins Nagasaki-1, Nagasaki-2, Nagasaki-3, and Hiroshima-1. Drs. J. Fine (Paris, France) and M. Galliano (Pavia, Italy) provided proalbumins Lille and Pollibauer. Dr. S. Migita (Kanazawa, Japan) provided albumin Takefu. Dr. C.-Y. Chuang (Taipei, Taiwan) provided albumin Taipei. Dr. T. Lau (Gainesville, FL) provided albumin Gainesville. Dr. T. Peters, Jr. (Cooperstown, NY) provided albumin Westcott. Dr. L. Weitkamp (Rochester, NY) provided several specimens of albumin B.

Sera from individuals with an albumin genetic variant were screened electrophoretically. None of the eight specimens studied was from individuals known to be closely related. The phenotypes studied were Mexico-2 (Me-2) both homozygous (Me-2/Me-2) (sera C276683 and C276685) and heterozygous (A/Me-2) (sera C276681 and C276682) from the Pima population in Arizona, heterozygous Naskapi (A/Na) (serum C276684) from the Pima population, homozygous Naskapi (Na/Na) (serum C12765) from the Montagnais population in Canada, and heterozygous Mersin (A/Mersin) (serum C180460) and Adana (A/Adana) (serum C180458) found in Eti Turks. As a reference standard the researchers used a commercial human albumin supposed to be A/A (lot 102578, Calbiochem-Behring, La Jolla, CA).

Study

Tokunaga, K., Ohashi, J., Bannai, M. and Juji, T. 2001. Genetic Link Between Asians and American Indians: Evidence from HLA Genes and Haplotypes. Human Immunology, 62: 1001-1008.

Study Summary

The researchers have been studying polymorphisms of HLA class I and II genes in East Asians including Buryat in Siberia, Mongolian, Han Chinese, Man Chinese, Korean Chinese, South Korean, and Taiwan indigenous population in collaboration with many Asian scientists. Regional populations in Japan, Hondo-Japanese, Ryukyuan, and Ainu, were also studied. HLA-A, -B, and -DRB1 gene frequencies were subjected to the correspondence analysis and calculation of DA distances. The correspondence analysis demonstrated several major clusters of human populations in the world. "Mongoloid" populations were highly diversified, in which several clusters such as Northeast Asians, Southeast Asians, Oceanians, and American Indians were observed. Interestingly, an indigenous population in North Japan, Ainu, was placed relatively close to American Indians in the correspondence analysis. Distribution of particular HLA-A, -B, -DRB1 alleles and haplotypes was also analyzed in relation to migration and dispersal routes of ancestral populations. A number of alleles and haplotypes showed characteristic patterns of regional distribution. For example, B39-HR5-DQ7 (B*3901-DRB1*1406-DQB1*0301) was shared by Ainu and American Indians. A24-Cw8-B48 was commonly observed in Taiwan indigenous populations, Maori in New Zealand, Orochon in Northeast China, Inuit, and Tlingit. These findings further support the genetic link between East Asians and American Indians. They have proposed that various ancestral populations in East Asia, marked by different HLA haplotypes, had migrated and dispersed through multiple routes. Moreover, relatively small genetic distances and the sharing of several HLA haplotypes between Ainu and American Indians suggest that these populations are descendants of some Upper Paleolithic populations of East Asia.

Genetic Materials and Lines

Regional populations in Japan, Hondo-Japanese in major islands, Ryukyuan in South Japan, and Ainu in North Japan, were studied. For the other populations, the researchers referred to previous reports. HLA data of Senegalese, South

Africans, San, Khoi, North American Africans, Danes, French, Germans, Italians, Romanians, Sardinians, Spanish, Spanish Gypsies, Asian Indians, Non-Austronesian Highlanders in Papua New Guinea, Buyi in South China, Singapore Chinese, and Tlingit in Alaska were obtained from the summary report at the 11th International Histocompatibility Workshop (IHW). The data of Kazakh in West China were obtained from the researchers joint report presented at the 12th IHW. The data of Bubi in island of Bioko (Equatorial Guinea), Zulu in South Africa, Moroccan and Ashkenazi Jews, Greeks, Croatians, Orcadians, Australian aboriginal Wailbri, Bunun in Taiwan, Dai Dam, and Dai Lue in Thai, urban Thais, Taiwanese, Khoton and Khalkh Mongolians, Nivkhi in Sakhalin (Far Eastern Russia), two Guaicuruan-speaking Argentinian tribes, Toba and Wichi, three Brazilian tribes, Kaingang, Guarani, and Terena, Bari Amerindians living on the border between Venezuela and Colombia, and Seris in Mexico were obtained from previous reports.

Study

Torroni, A., Schurr, T. G., Cabell, M. F., Brown, M. D., Neel, J. V., Larsen, M., Smith, D. G., Vullo, C. M. and Wallace, D. C. 1993. Asian affinities and continental radiation of the four founding American Indian mtDNAs. American Journal Of Human Genetics, 53(3): 563-590.

Study Summary

The mtDNA variation of 321 individuals from 17 American Indian populations was examined by high-resolution restriction endonuclease analysis. All mtDNAs were amplified from a variety of sources by using PCR. The mtDNA of a subset of 38 of these individuals was also analyzed by D-loop sequencing. The resulting data were combined with previous mtDNA data from five other American Indian tribes, as well as with data from a variety of Asian populations, and were used to deduce the phylogenetic relationships between mtDNAs and to estimate sequence divergences. This analysis revealed the presence of four haplotype groups (haplogroups A, B, C, and D) in the Amerind, but only one haplogroup (A) in the Na-

Dene, and confirmed the independent origins of the Amerinds and the Na-Dene. Further, each haplogroup appeared to have been founded by a single mtDNA haplotype, a result that is consistent with a hypothesized founder effect. Most of the variation within haplogroups was tribal specific, that is, it occurred as tribal private polymorphisms. These observations suggest that the process of tribalization began early in the history of the Amerinds, with relatively little intertribal genetic exchange occurring subsequently. The sequencing of 341 nucleotides in the mtDNA D-loop revealed that the D-loop sequence variation correlated strongly with the four haplogroups defined by restriction analysis, and it indicated that the D-loop variation, like the haplotype variation, arose predominantly after the migration of the ancestral Amerinds across the Bering land bridge.

Genetic Materials and Lines

Drs. R.M. Fourney and C. Fregeau of the Royal Canadian Mounted Police donated the Ojibwa DNA samples from northern Ontario. The 361 samples of blood cells analyzed were a random selection from a much larger collection of American Indian samples. They were obtained from 311 Amerinds and 50 Na-Dene. The locations of the 17 Amerind and 2 Na-Dene tribes analyzed in this study are presented in figure 1, along with those of the 3 Amerind and 2 Na-Dene tribes, which were previously analyzed (Torroni et al. 1992).

Study

Torroni, A., and Wallace, D.C. 1995. mtDNA Haplogroups in Native Americans. American Journal of Human Genetics, 56(5): 1234-1238.

Study Summary
Not Available.

Genetic Materials and Lines
Not Available.

Study

Torroni, A., Schurr, T. G., Yang, C. C., Szathmary, E. J., Williams, R. C., Schanfield, M. S., Troup, G. A., Knowler, W. C., Lawrence, D. N. and Weiss et, a. 1992. Native American mitochondrial DNA analysis indicates that the Amerind and the Nadene populations were founded by two independent migrations. Genetics, 130(1): 153-162.

Study Summary

Mitochondrial DNAs (mtDNAs) from 167 American Indians including 87 Amerind-speakers (Amerinds) and 80 Nadene-speakers (Nadene) were surveyed for sequence variation by detailed restriction analysis. All American Indian mtDNAs clustered into one of four distinct lineages, defined by the restriction site variants: HincII site loss at np 13,259, AluI site loss at np 5,176, 9-base pair (9-bp) COII-tRNA(Lys) intergenic deletion and HaeIII site gain at np 663. The HincII np 13,259 and AluI np 5,176 lineages were observed exclusively in Amerinds and were shared by all such tribal groups analyzed, thus demonstrating that North, Central and South American Amerinds originated from a common ancestral genetic stock. The 9-bp deletion and HaeIII np 663 lineages were found in both the Amerinds and Nadene but the Nadene HaeIII np 663 lineage had a unique sublineage defined by an RsaI site loss at np 16,329. The amount of sequence variation accumulated in the Amerind HincII np 13,259 and AluI np 5,176 lineages and that in the Amerind portion of the HaeIII np 663 lineage all gave divergence times in the order of 20,000 years before present. The divergence time for the Nadene portion of the HaeIII np 663 lineage was about 6,000-10,000 years. Hence, the ancestral Nadene migrated from Asia independently and considerably more recently than the progenitors of the Amerinds. The divergence times of both the Amerind and Nadene branches of the COII-tRNA(Lys) deletion lineage were intermediate between the Amerind and Nadene specific lineages, raising the possibility of a third source of mtDNA in American Indians.

Genetic Materials and Lines

C. Ronald Scott provided the Tlingit samples. Blood

cells in the form of lymphoblasts, platelets, or buffy coats were obtained from 167 subjects including 80 Nadene and 87 Amerinds. The Nadene were comprised of 30 Dogrib (NW Canada), 48 Navajo (Arizona and New Mexico) and 2 Tlingits (Alaska), and their bloods represent a random sampling of individuals within each respective tribal group. The Amerinds were comprised of 30 Pima (Arizona), 1 Hopi (Arizona), 1 Pomo (California), 27 Maya (Mexico) and 28 Ticuna (Brazil). The Pima and Ticuna individuals from whom blood was taken were known to be unrelated for at least three generations, whereas the Maya represent a random sample of that tribal group. The Pima was 1.0% (WILLIAMS et al. 1985, 1986). Bloods were collected at the Gila River Indian Community in Sacaton, Arizona, by the NIDDK as part of a diabetes study (KNOWLER et al. 1978).

Study

Torroni, A., Sukernik, R. I., Schurr, T. G., Starikovskaya, Y. B., Cabell, M. F., Crawford, M. H., Comuzzie, A. G. and Wallace, D. C. 1993. Mtdna Variation of Aboriginal Siberians Reveals Distinct Genetic Affinities with Native-Americans. American Journal of Human Genetics, 53(3): 591-608.

Study Summary

The mtDNA variation of 411 individuals from 10 aboriginal Siberian populations was analyzed in an effort to delineate the relationships between Siberian and American Indian populations. All mtDNAs were characterized by PCR amplification and restriction analysis, and a subset of them was characterized by control region sequencing. The resulting data were then compiled with previous mtDNA data from American Indians and Asians and were used for phylogenetic analyses and sequence divergence estimations. Aboriginal Siberian populations exhibited mtDNAs from three (A, C, and D) of the four haplogroups observed in American Indians. However, none of the Siberian populations showed mtDNAs from the fourth haplogroup, group B. The presence of group B deletion haplotypes in East Asian and American Indian populations but their absence in Siberians raises the possibility that haplogroup B could represent a migratory event distinct from the one(s) which brought group A, C, and D mtDNAs to the Americas. The researchers findings support the hypothesis that the first humans to move from Siberia to the Americas carried with them a limited number of founding mtDNAs and that the initial migration occurred between 17,000-34,000 years before present.

Genetic Materials and Lines

A total of 441 individuals from 10 aboriginal populations of northern Siberia and the Russian Far East were selected for mtDNA characterization. These samples include 20 Sel'kups, 49 Nganasans, 43 Evens, 51 Evenks, 46 Udegeys, 24 Chukchi, 46 Koryaks, 27 Yukagirs, 57 Nivkhs, and 50 Asiatic Eskimos.

Study

Underhill, P. A., Jin, L., Zemans, R., Oefner, P. J. and Cavalli-Sforza, L. L. 1996. A pre-Columbian Y chromosome-specific transition and its implications for human evolutionary history. Proceedings Of The National Academy Of Sciences Of The United States Of America, 93(1): 196-200.

Study Summary

A polymorphic CàT transition located on the human Y chromosome was found by the systematic comparative sequencing of Y-specific sequence-tagged sites by denaturing high-performance liquid chromatography. The results of genotyping representative global indigenous populations indicate that the locus is polymorphic exclusively within the Western Hemisphere. The pre-Columbian T allele occurs at > 90% frequency within the native South and Central American populations examined, while its occurrence in North America is approximately 50%. Concomitant genotyping at the polymorphic tetranucleotide microsatellite DYS19 locus revealed that the CàT mutation displayed significant linkage disequilibrium with the 186-bp allele. The data suggest a single origin of linguistically diverse American Indians with subsequent haplotype differentiation within radiating indigenous populations as well as post-Columbian European and African gene flow. The mutation may have originated either in North America at a very early time during the expansion or before it, in the ancestral population(s) from which all Americans may have originated. The analysis of linkage of the DYS199 and the DYS19 tetranucleotide loci suggests that the C—>T mutation may have occurred around 30,000 years ago. The researchers estimate the nucleotide diversity over 4.2 kb of the nonrecombining portion of the Y chromosome to be 0.00014. compared to autosomes, the majority of variation is due to the smaller effective population size of the Y chromosome rather than selective sweeps. There begins to emerge a pattern of pronounced geographical localization of Y-specific nucleotide substitution polymorphisms.

Genetic Materials and Lines

The Eskimo samples were provided by Mary Ann Walkinshaw, Department of Public Safety, Alaska, while Navajo DNA was donated by Gary Troup, University of New Mexico School of Medicine.

Study

Underhill, P. A., Passarino, G., Lin, A. A., Shen, P., Lahr, M. M., Foley, R. A., Oefner, P. J. and Cavalli-Sforza, L. L. 2001. The Phylogeography of Y Chromosome Binary Haplotypes and the Origins of Modern Human Populations. Annuals of Human Genetics, 65: 43-62.

Study Summary

Although molecular genetic evidence continues to accumulate that is consistent with a recent common African ancestry of modern humans, its ability to illuminate regional histories remains incomplete. A set of unique event polymorphisms associated with the non-recombining portion of the Y-chromosome (NRY) addresses this issue by providing evidence concerning successful migrations originating from Africa, which can be interpreted as subsequent colonizations, differentiations and migrations overlaid upon previous population ranges. A total of 205 markers identified by denaturing high performance liquid chromatography (DHPLC), together with 13 taken from the literature, were used to construct a parsimonious genealogy. Ancestral allelic states were deduced from orthologous great ape sequences. A total of 131 unique haplotypes were defined which trace the microevolutionary trajectory of global modern human genetic diversification. The genealogy provides a detailed phylogeographic portrait of contemporary global population structure that is emblematic of human origins, divergence and population history that is consistent with climatic, paleoanthropological and other genetic knowledge.

Genetic Materials and Lines

DNA from 1062 men belonging to 21 populations was analyzed. Further details on the ethnic affiliations of these samples are given in Underhill et al. (2000).

Study

Villalobos-Arambula, A. R., Rivas, F., Sandoval, L., Perea, F. J., Casas-Castaneda, M., Cantu, J. M. and Ibara, B. 2000. ?^{A} Globin Gene Haplotypes in Mexican Huichols: Genetic Relatedness to Other Populations. American Journal of Human Biology, 12: 201-206.

Study Summary

The haplotypes of 97 ?^{A} independent chromosomes from a Mexican Huichol American Indian group were analyzed. The analysis also included 87 ?^{A} chromosomes from a Mexican Mestizo population previously studied. Among Huichols, eight different 58 b haplotypes (5Hps) were observed, with types 1(+ - - - -), 13(+ + + - +) and 2(- + + - +) at frequencies of 0.794, 0.093, and 0.041, respectively. In Mestizos, 17 5Hps were found, types 1, 3(- + - + +), 2, 5(- + - - +) and 9(- - - - -) being the most common at frequencies of 0.391, 0.172, 0.092, 0.069, and 0.046, respectively. 38 haplotype (3Hps) frequency distributions were 0.443(+ +), 0.083(+ -), and 0.474(-+) in Huichols and 0.563(+ +), 0.149(+ -), and

0.287(- +) in Mestizos. Pairwise comparison for both haplotype distributions between the two populations showed significant differences. Pairwise distributions of 3Hps for Huichols were compared with nine worldwide populations, three African, two Asian, two Melanesian, one Caucasian, and one United States American Indian. The distributions of the Huichol were different ($P < 0.05$) from all populations except the American Indian. Nei's genetic distances showed the Huichols to be closer to the American Indians, followed by Melanesians from Vanuatu and Asians; Africans were the farthest. The 5Hp distributions in Mexicans were also compared with 23 worldwide populations (including African, American Indian, Asian, Caucasian, and Pacific Islanders). Huichol distributions were different ($P < 0.05$) from all other populations except Koreans. The Mestizo distribution was also different from the others, except three Caucasian groups. Nei's genetic distance between the same populations disclosed that the Huichols are in relatively close proximity to five out of six Asian populations considered. The same analysis with grouped worldwide populations showed American Indians as population closest to the Huichols, followed by Pacific Islanders and Asians. Present observations are consistent with an important Asian contribution to the Huichol genome in this chromosomal region.

Genetic Materials and Lines

Blood samples were collected from 53 Huichol Indians, 50 of whom were unrelated. They were from the Mezquitic (42) and the Nayar regions (11).

Study

Wallace, D.C., Garrison, K., and Knowler, W.C. 1985. Dramatic Founder Effects in Amerindian Mitochondrial DNAs. American Journal of Physical Anthropology, 68(2): 149-155.

Study Summary

Southwestern American Indian (Amerindian) mitochondrial DNAs (mtDNAs) were analyzed with restriction endonucleases and found to contain Asian restriction fragment length polymorphisms (RFLPs) but at frequencies very different from those found in Asia. One rare Asian HincII RFLP was found in 40% of the Amerindians. Several mtDNAs were discovered which have not yet been observed on other continents and different tribes were found to have distinctive mtDNAs. Since the mtDNA is inherited exclusively through the maternal lineage, these results suggest that Amerindian tribes were founded by small numbers of female lineages and that new mutations have been fixed in these lineages since their separation from Asia.

Genetic Materials and Lines
Not Available.

Study
Ward, R. H., Alan Redd, A., Valencia, D., Frazier, B. and Paabo, S. 1993. Genetic and Linguistic Differentiation in the Americas. Proceedings of the National Academy of Sciences, USA, 90: 10663-10667.

Study Summary
The relationship between linguistic differentiation and evolutionary affinities was evaluated in three tribes of the Pacific Northwest. Two tribes (Nuu-Chah-Nulth and Bella Coola) speak Amerind languages, while the language of the third (Haida) belongs to a different linguistic phylum- Na-Dene. Construction of a molecular phylogeny gave no evidence of clustering by linguistic affiliation, suggesting a relatively recent ancestry of these linguistically divergent populations. When the evolutionary affinities of the tribes were evaluated in terms of mitochondrial sequence diversity, the Na-Dene-speaking Haida had a reduced among of diversity compared to the two Amerind tribes and thus appear to be a biologically younger population. Further, since the sequence diversity between the two Amerind-speaking tribes is comparable to the diversity between the Amerind tribes and the Na-Dene Haida, the evolutionary divergence within the Amerind linguistic phylum may be as great as the evolutionary divergence

between the Amerind and Na-Dene phyla. Hence, in the New World, rates of linguistic differentiation appear to be markedly faster than rates of biological differentiation, with little congruence between linguistic hierarchy and the pattern of evolutionary relationships.

Genetic Materials and Lines

Dr. Philip Gofton provided access to the archival serum samples. The 41 individuals, who were originally sampled as part of a rheumatic disease survey, represent both Queen Charlotte communities and claimed genealogical descent from Haida individuals believed to have lived on the Queen Charlottes during the late 19th century. The 40 Bella Coola individuals in this study were also randomly selected from patients originally surveyed for rheumatic disease.

Study

Ward, R. H., Frazier, B. L., Dew-Jager, K. and Paabo, S. 1991. Extensive Mitochondrial Diversity within a Single Amerindian Tribe. Proceedings of the National Academy of Sciences, USA, 88: 8720-8724.

Study Summary

Sequencing of a 360-nucleotide segment of the mitochondrial control region for 63 individuals from an Amerindian tribe, the Nuu-Chah-Nulth of the Pacific Northwest, revealed the existence of 28 lineages defined by 26 variable positions. This represents a substantial level of mitochondrial diversity for a small local population. Furthermore, the sequence diversity among these Nuu-Chah-Nulth lineages is >60% of the mitochondrial sequence diversity observed in major ethnic groups such as Japanese or sub-Saharan Africans. It was also observed that the majority of the mitochondrial lineages of the Nuu-Chah-Nulth fell into phylogenetic clusters. The magnitude of the sequence difference between the lineage clusters suggests that their origin predates the entry of humans into the Americas. Since a single Amerindian tribe can contain such extensive molecular diversity, it is unnecessary to presume that substantial genetic bottlenecks occurred during the formation of contemporary ethnic groups. In particular, these data do not support the concept of a dramatic founder effect during the peopling of the Americas.

Genetic Materials and Lines

The Nuu-Chah-Nulth collaborated with this study. As part of a biomedical study, the traditional band communities, numbering some 2000-2400 people, were surveyed between 1984 and 1986. Serum samples were collected from a large proportion (45%) of the population, and detailed genealogical information was collected for each band, along with basic demographic data. To determine the amount of mitochondrial variability, the researchers selected 63 maternally unrelated individuals whose genealogy indicated Nuu-Chah-Nulth descent at least as far back as the late nineteenth century. These individuals were selected from 13 of the 14 contemporary bands. An additional 5 individuals, each known to be maternally related to 1 of the 63 independent individuals, were also included in the study as positive controls.

Study

Williams, R. C., Knowler, W. C., Pettitt, D. J., Long, J. C., Rokala, D. A., Polesky, H. F., Hackenberg, R. A., Steinberg, A. G. and Bennett, P. H. 1992. The Magnitude and Origin of European-American Admixture in the Gila River Indian Community of Arizona - a Union of Genetics and Demography. American Journal of Human Genetics, 51(1): 101-110.

Study Summary

Complementary genetic and demographic analyses estimate the total proportion of European-American admixture in the Gila River Indian Community and trace its mode of entry. Among the 9,616 residents in the sample, 2,015 persons claim only partial American Indian heritage. A procedure employing 23 alleles or haplotypes at eight loci was used to estimate the proportion of European-American admixture, $m(a)$, for the entire sample and within six categories of Caucasian admixture calculated from demographic data, $m(d)$.

The genetic analysis gave an estimate of total European-American admixture in the community of 0.054 (95% confidence interval [CI] .044-.063), while an estimate from demographic records was similar, .059. Regression of m(a) on m(d) yielded a fitted line m(a) = .922m(d), r = .959 (P = .0001). When total European-American admixture is partitioned between the contributing populations, Mexican-Americans have provided .671, European-Americans .305, and African-Americans .023. These results are discussed within the context of the ethnic composition of the Gila River Indian Community, the assumptions underlying the methods, and the potential that demographic data have for enriching genetic measurements of human admixture. It is concluded that, despite the severe assumptions of the mathematical methods, accurate, reliable estimates of genetic admixture are possible from allele and haplotype frequencies, even when there is little demographic information for the population.

Genetic Materials and Lines

The Gila River Indian Community cooperated in this study. The staff of the Diabetes and Arthritis Epidemiology Section, NIDDK, collected the demographic data and conducted the examinations. Blood samples were obtained and tested for a range of genetic markers. Allotypes for the Gm and Km systems were determined in the laboratory of A.G.S., while the Rhesus, ABO, MNSs, Duffy, Kidd, and Kell systems were typed by the Memorial Blood Center of Minneapolis. Standard methods have been used for all assays. Frequencies for the following alleles or haplotypes have been included in the present study: (Rhesus) CDe, cDE, CDE, and cde; (ABO) A_1, A_2, B, and O; (MNSs) MS, Ms, NS, and Ns; (Duffy) Fy^a and Fy^b; (Kidd) Jk^a and Jk^b; (Kell) K and k; (Gm) $Gm^{1;21}$, $Gm^{1,2;21}$, and $Gm^{3;5,13,14}$; and (Km) Km^1 and Km^3. Allele frequencies for the Caucasian parental group came from paternity cases and were taken from tables of large samples published by the American Association of Blood Banks.

Study

Williams, S. R., Chagnon, N. A. and Spielman, R. S. 2002. Nuclear and mitochondrial genetic variation in the Yanomamo: a test case for ancient DNA studies of prehistoric populations. American Journal of Physical Anthropology, 117(3): 246-59.

Study Summary

Ancient DNA provides a potentially revolutionary way to study biological relationships in prehistoric populations, but genetic patterns are complex and require careful interpretation based on robust, well-tested models. In this study, nuclear and mitochondrial markers were compared in the Yanomamo, to assess how well each data set could differentiate among closely related groups. The villages selected for the study share a recent fission history and are closely related to each other, as would likely be the case among prehistoric peoples living in the same valley or region. The Yanomamo generally practice village-level endogamy, but some migration and gene flow are known to occur between villages. Nuclear and mitochondrial DNA data were compared using F-statistics and genetic distance analyses. The nuclear data performed as expected, males and females from the same village were similar, and the villages were genetically distinct, with the magnitude of genetic differences correlated with historical relationship. However, mtDNA analyses did not yield the expected results. The genetic distances between villages did not correlate with historical relationship, and the sexes were significantly different from each other in two villages. Both the Lane and Sublett and the Spence methods, used to test for archaeological residence patterns, were consistent with endogamy. Hence, ancient DNA can, in principle, provide us with a unique opportunity to study genetic structure and gene flow in archaeological populations. However, interpretations, particularly those based on single loci such as mitochondrial DNA, should be cautious because sex-specific migration and sampling issues may have dramatic effects.

Genetic Materials and Lines

Permission to use Yanomamo samples collected by

Neel and colleagues was given by K.M. Weiss (Penn State University). P. Ladenson and R. Cooper (Johns Hopkins University) granted us permission to use Washawa blood samples.

Total genomic DNA was isolated from 173 individuals, with 136 from red blood cell pellets collected by Neel et al., and 37 from whole blood samples collected by Ladenson and Cooper. The majority of red blood cell pellet samples (n = 119) were selected because they could be located on pedigrees of individuals collected by Neel et al. in 1968 and 1972. Seventeen additional samples collected by Neel et al. were included in order to augment the village sample sizes, but could not be located on the pedigrees. Thirty-seven Village 5 individuals were included from a more recent expedition made by Paul Ladenson, David S. Cooper, and Chagnon in 1992. Residence pattern studies require that samples include only postmigration individuals, and so individuals of adult or marriageable age were chosen. Village samples were divided nearly equally by sex.

Study

Zegura, S. L., Karafet, T. M., Zhivotovsky, L. A. and Hammer, M. F. 2004. High-resolution SNPs and microsatellite haplotypes point to a single, recent entry of American Indian Y chromosomes into the Americas. Molecular Biology and Evolution, 21(1): 164-175.

Study Summary

A total of 63 binary polymorphisms and 10 short tandem repeats (STRs) were genotyped on a sample of 2,344 Y chromosomes from 18 American Indian, 28 Asian, and 5 European populations to investigate the origin(s) of American Indian paternal lineages. All three of Greenberg's major linguistic divisions (including 342 Amerind speakers, 186 Na-Dene speakers, and 60 Aleut-Eskimo speakers) were represented in the researchers sample of 588 American Indians. Single-nucleotide polymorphism (SNP) analysis indicated that three major haplogroups, denoted as C, Q, and R, accounted for nearly 96% of American Indian Y chromosomes. Haplogroups

C and Q were deemed to represent early American Indian founding Y chromosome lineages; however, most haplogroup R lineages present in American Indians most likely came from recent admixture with Europeans. Although different phylogeographic and STR diversity patterns for the two major founding haplogroups previously led to the inference that they were carried from Asia to the Americas separately, the hypothesis of a single migration of a polymorphic founding population better fits the researchers expanded database. Phylogenetic analyses of STR variation within haplogroups C and Q traced both lineages to a probable ancestral homeland in the vicinity of the Altai Mountains in Southwest Siberia. Divergence dates between the Altai plus North Asians versus the American Indian population system ranged from 10,100 to 17,200 years for all lineages, precluding a very early entry into the Americas.

Genetic Materials and Lines

The researchers analyzed 63 binary single-nucleotide polymorphisms (SNPs) and 10 short tandem repeats (STRs) on a sample of 2,344 Y chromosomes from 51 populations representing the Americas, Asia, and Europe. The American Indian sample included 588 individuals from 18 populations allocated to Greenberg's (1987) three major American Indian language families as follows: 342 Amerind speakers, 186 Na-Dene speakers, and 60 Aleut-Eskimo speakers. American Indian linguistic affiliations, sample sizes, SNP haplogroup frequencies and diversity measures, STR haplotype numbers and repeat number variances, and three-letter and numerical population codes are given, which also contains summary genetic data for six geographically defined Eurasian groupings (Europe, North Asia, Central Asia, South Asia, East Asia, and Southeast Asia). Many of the individuals analyzed here were included in the researchers previous studies (Karafet et al. 1997, 1999, 2001, 2002; Hammer et al. 2001); however, the researchers most recent New World publication (Karafet et al. 1999) contained data from only 12 biallelic polymorphisms and two STRs on a total of 380 American Indians. New samples collected for this study came from the Apache, Navajo, Sioux, and Maya. All sampling protocols were approved by the

Human Subjects Committee at the University of Arizona.

Study

Zheng, H. Y., Sugimoto, C., Hasegawa, M., Kobayashi, N., Kanayama, A., Rodas, A., Mejia, M., Nakamichi, J., Guo, J., Kitamura, T. and Yogo, Y. 2003. Phylogenetic relationships among JC virus strains in Japanese/Koreans and American Indians speaking Amerind or Na-Dene. Journal of Molecular Evolution, 56(1): 18-27.

Study Summary

Many genetic studies using human mtDNA or the Y chromosome have been conducted to elucidate the relationships among the three American Indian groups speaking Amerind, Na-Dene, and Eskimo-Aleut. Human polyomavirus JC (JCV) may also help to gain insights into this issue. JCV isolates are classified into more than 10 geographically distinct genotypes (designated subtypes here), which were generated by splits in the three superclusters, Types A, B, and C. A particular subtype of JCV (named MY) belonging to Type B is spread in both Japanese/Koreans and American Indians speaking Amerind or Na-Dene. In this study, the researchers evaluated the phylogenetic relationships among MY isolates worldwide, using the whole-genome approach, with which a highly reliable phylogeny of JCV isolates can be reconstructed. Thirty-six complete sequences belonging to MY (10 from Japanese/Koreans, 24 from American Indians, and 2 from others), together with 54 belonging to other subtypes around the world, were aligned and subjected to phylogenetic analysis using the neighbor-joining and maximum-likelihood methods. In the resultant phylogenetic trees, the MY sequences diverged into two Japanese/Korean and five American Indian clades with high bootstrap probabilities. Two of the American Indian clades contained isolates mainly from Na-Denes and the others contained isolates mainly from Amerinds. The Na-Dene clades were not clustered together, nor were the Amerind clades. In contrast, the two Japanese/Korean clades were clustered at a high bootstrap probability. The researchers concluded that there is no distinction between Amerinds and Na-Denes in terms of indigenous JCVs, although they are linguistically distinguished from each other.

Genetic Materials and Lines

The researchers collected urine samples from American Indians living in Guatemala, Peru, Mexico, and Canada. The Guatemalan urine donors were native patients in hospitals located at Totonicapan, Santa Cruz del Quiche, and Solola, Guatemala. The Peruvian donors were native inhabitants of villages located in Ollantay Tambo and Chuquito. The Mexican and Canadian urine donors were volunteers living in the state of Chihuahua, Mexico, and the northern part of Alberta, Canada, respectively. All urine donors were over 40 years old. The urine donors in Mexico, Guatemala, and Peru belonged to various subgroups of the Amerind language group, and those in Canada belonged to the Na-Dene language group.

INDEX

Ache 117
adenine 7
aDNA 1, 16, 47, 60-64, 124
Africa 53, 70
African 48, 58, 108-110, 119, 121-122, 125-127, 134, 136, 158, 164-165, 168-169, 175, 188, 192, 206, 208, 212
Ainu 48
Aleut-Eskimo 107, 214-215
Aleuts 74
allele 15, 27, 29, 31
allele frequencies 10, 13, 22, 27, 29
alleles 8, 12-13
Altai 115, 129, 159, 176, 180, 189-190, 215
American Indian 4, 23-24
American Indians 15
Amerind 56, 114-115, 117, 150, 152, 161-162, 178-179, 189, 201-203, 209-210, 214-217
Amerindian 55, 57, 60, 62, 65
Amerindians 107, 109, 111-112, 144, 159-160, 166-167, 171-172, 180-181, 183, 201, 209
Anasazi 123, 170
anatomically modern humans 5
ancestral polymorphism 21-22
ancient DNA 1
Apache 127, 134-135, 176, 215
Arara 113, 192
archaeology 4
Asia 23-24, 29, 68, 72-73
Asian populations 47-49, 53
Asurini 117
Atacameno 55
Athapascan 113
Athapaskan 115, 123, 177, 190

Awa-Guaja 113
Barira 117
Bella Coola 113, 115, 153-154, 189, 209-210
Bering Land Bridge 24
Bering landbridge 135
Beringia 114-115, 148, 183, 189, 191, 194
bi-allelic 12
biological affiliation 4-5, 16, 21-22, 24, 28
bootstrap method 48
bottleneck 29
Brazil 111, 116-117, 130, 134, 136, 182, 192, 204
Buryad 118
Buryats 128-129, 148, 180
Caucasian 108-110, 120-121, 163, 168-169, 208, 211-212
Cavalli-Sforza 3, 5, 39-41, 45
Celilo Falls 36
Cheyenne 60, 140, 156
Chinchorro 64
Chinese 109-110, 119, 127-128, 140, 180, 200-201
Chipewayan 60
chromosomes 7-10
Chukchi 48, 114-115, 139, 148, 176-177, 189-190, 194-195, 205
Cinta-Larga 118
Clovis 68, 74, 113, 183
coalescence 9
coalescence theory 20, 23
coalescence times 68
coalescent times 22, 24-25
coalescent tree 19
Colorado 38
control region 15
cultural integration 34
cytidine 7
deoxyribonucleic acid 1, 7
Durvud 118
Egyptian mummy 61
endogamous 34
Eskimo 107, 113-115, 133, 150, 152, 159, 162-163, 176-177, 183, 190, 194-195, 206, 214-216

Eskimo-Aleut 56
Eskimos 74
ethnographic record 34, 36-37
Europeans 48, 53-54
exogamous 34, 37
Florida 52
Folo 58
founder effect 41, 45
Fremont 61, 63, 124, 171-172
Gaviao 113-115
Gene flow 43-44
gene frequencies 28, 31
genetic bottleneck 107
genetic drift 35, 37, 41-43
geographic distance 35, 39-40
Gorotire 118
Great Basin 36, 61-63, 123-124, 138-139, 154
Greenland Eskimo 113, 115, 190
guanine 7
Guarani 118, 192, 201
Guaymi 111
Haida 113, 115, 176, 189, 209-210
Hakas 58
haplogroup 9, 13, 29, 67-70, 72, 74
haplogroups 15-16, 49, 52-57, 59-60, 62-65
haplotype 67, 69-70
haplotype frequency 16
haplotypes 9, 13, 15-16, 19, 29, 49, 52-53, 55-57, 59-60, 65
Havasupai 140
historic population movement 3
historic population movements 8-10, 12, 29, 31, 45, 51, 57, 60
hitch-hiking 16
HLA system 50
Hokan 131, 149
horse 38
Huetar 113-115
Huilliches 110
Huitoto 118
hunter-gatherers 35, 39

hypervariable segments 12
Illinois 62
inbreeding coefficient 20, 36, 40
Inca 64
Ingano 117, 179
Inuit Eskimo 30
Inuits 58
Inupiaq Eskimo 113, 115
island model 35-39, 41-42
isolation by distance 37, 39
Jamaican 108, 127
Japanese 58
Kaigang 118
Katuena 113, 192
Kaxuyana 111
Kayapo 111, 113, 192
Kennewick Man 4
Khalkh 118, 201
kinship coefficient 40
Koreans 58
Kraho 111, 118
Kuna 113, 115, 143
Last Glacial Maximum 72, 74
Lengua 110, 119
Mackenzie 39
Malecot 39-41
Mapuche 113-115, 119
Mapuches 110
maximum likelihood 19, 21
Maya 52, 64, 119, 134, 181-182, 204, 215
Mbuty Pygmy 115
Mekranoti 118
Mexico 121-122, 124-126, 140, 148-149, 151-153, 198-199, 201, 204, 206, 217
microsatellite 108, 110-111, 116-117, 147, 179, 188, 205, 214
microsatellites 12-13, 115, 119, 142, 178
migration 30-31, 33-37, 39, 41-44, 67, 69
migration rate 36-37, 44
migrations 1, 3-4

minisatellites 12-13
Mississippi 38
mitochondrial DNA 1, 9
mitosis 8
molecular anthropology 1-4, 7-9, 33, 45
molecular geneticists 1
Mongolia 115, 159-161
Mongolians 54
mtDNA 1-5, 15-16, 22, 27-31, 47, 49, 51-55, 57, 60, 62-64, 67-69, 73-74
mutation 10-11, 13, 19-20, 22-23, 35, 41-42
mutation rates 70, 73
mutations 8-13, 16-17
Na-Dene 56-57, 60, 74, 107, 110, 114-115, 120, 133, 148-150, 152, 159-160, 165-167, 175-176, 180, 183, 201-202, 209-210, 214-217
NAGPRA 4
Navajo 30, 60, 111-112, 119-121, 140, 176, 204, 206, 215
Navajos 58, 140, 148
Nevada 138-139
New Mexico 121, 140, 151, 204, 206
New World 107, 110-114, 128-131, 135-136, 139-140, 142-143, 148, 150, 152-153, 160-163, 165, 175-179, 182, 189, 191, 194, 197, 210, 215
Ngobe 60, 113-115, 143
nonbinomial 20
non-native admixture 56
Nootka 54
North American Indian 4
Northeast Asians 48
Northern Plains 36
Northwest Coast 36
nuclear DNA 3
nucleotide diversity 20-21
Numic 138-139
Nuu-Chah-Nulth 54, 113, 115, 120-121, 154, 189, 209-211
Ojibwa 120-121, 153, 156, 176, 187, 193, 202
Oneota 62-63
oral history 4
orthologous sequences 21

out-of-Africa 118
Paacas Novos 118
Paiute 62
Paleoindians 107
panmixis 36
Parakana 118
parent-offspring 38, 40
Pehuenches 110
Penutian 131, 166
Peopling of North America 1-2, 33-35, 37, 41, 43-44
Peopling of the Americas 113-114, 134, 139, 160, 176, 182, 211
physical anthropology 8
Pima 52, 55, 140, 181-182, 198-199, 204
Plains 36, 38
Plateau 36, 38
polymorphism 8, 15-16
polymorphisms 51, 55, 57-59
population density 33-34
Poturujara 113, 192
Pueblo 111-112, 123, 151
Puebloan 63
Pyramid Lake 62, 139
random mating 20, 24, 36, 38
Recombination 10, 13
Reconciled trees 31
restriction fragment length polymorphism 15
RFLP 12, 15, 51
Rio Grande 38
Seminoles 111, 148
Sewall Wright 34
short tandem repeat 58
short tandem repeats 27
Shoshone 62
Siberia 24
Siberian 67, 69, 71-72
Siberian populations 47, 53-55, 58
single migratory wave 54
single nucleotide 58-59
single wave of migration 4

SNP 70-73
South American Indian 51, 60
South American Indians 5
Southwest 38
Spirit Cave Mummy 4
St. Lawrence 38
stepping-stone model 42
Stillwater Marsh 62, 139
STR 70, 73
Surui 113-114, 136
Tanana 60, 140, 190
Tehuelches 110
The Dalles 36
thymidine 7
Ticuna 52, 118, 179, 181-182, 204
Tiriyo 113
Tiryio 118
Tuva 129, 148, 160
Ulan Bator 54
Urubu-Kaapor 118
Utah 63
Wai Wai 113-114, 136
Waiapi 118
Wai-Wai 115, 136
Warao 117
Wayampi 113
Wayuu 118, 179
Wichis 110
Xavante 113-115, 136
Xikrin 118
Y chromosome 1-5, 8, 10-13, 15-16, 22, 24, 27-31, 47, 51, 57-60, 70-72, 74
Yagua 118
Yakima 113, 115, 120, 189-190
Yanomama 113, 158, 192, 199
Yanomami 113-114, 129-130, 158
Yoruba 115
Yukon 39
Yukpa 118
Zoro 113-115

www.ingramcontent.com/pod-product-compliance
Ingram Content Group UK Ltd.
Pitfield, Milton Keynes, MK11 3LW, UK
UKHW041228200426
11947UKWH00034B/371